HIS STORY for US

PUBLISHING

HOW MODERN COSMOLOGY TRANSFORMS CHRISTIAN UNDERSTANDING

The Physics of Faith

The Physics of Faith: How Modern Cosmology Transforms Christian Understanding

By Arthur A.Tiger.

A groundbreaking exploration of how modern physics—from dark matter to quantum mechanics—deepens rather than threatens Christian faith, revealing profound connections between cosmological discoveries and ancient theological wisdom.

Published by His Story for Us

www.arthurtigerbooks.com

First Edition: 2025

 The author has attempted to recreate events, locales, and conversations from memory and notes. In some instances, names and identifying details have been changed to protect privacy.

ISBN 978-2-6073-5115-0

Cosmic Wonder and Christian Faith

The Two Books and Their Tensions

In the thirteenth century, St. Bonaventure wrote of God's two books: Scripture and Creation. Today, as I stand beneath a night sky punctured by stars whose light left before humans walked upright, these two books sometimes seem to be written in different languages, telling different stories, pointing toward different conclusions.

The cosmos that modern science reveals staggers the imagination. We inhabit a universe 13.8 billion years old, composed of two trillion galaxies, each containing hundreds of billions of stars. Space itself expands, carrying galaxies apart like seeds scattered by an infinite wind. Most of what exists—95% of it—remains invisible to us, dark matter and dark energy we can detect only through their ghostly gravitational effects. And according to our best physics, this magnificent cosmos is headed toward heat death, a state of maximum entropy where nothing interesting will ever happen again, forever and ever, amen.

Meanwhile, in small buildings on Sunday mornings, people gather to sing about purpose and meaning, to pray to a God they believe crafted each human soul with intention, to read ancient texts that speak of divine providence and eternal hope. The disconnect can be jarring. How do we hold together the vastness that astronomy reveals with the intimacy that faith proclaims? Can we honestly maintain that humans matter in a universe where our entire galaxy is less than a rounding error?

Why This Conversation Matters

I think particularly of three groups of people who need this conversation, though they might not know they need it. First, there are the scientists who are Christians, living daily at the intersection of research and faith. I know a cosmologist who studies the cosmic microwave background on weekdays and teaches Sunday school on weekends. The cognitive dissonance is real. She doesn't compartmentalize as much as live in constant translation, moving between languages that don't quite map onto each other. She needs frameworks for integration that don't require intellectual compromise in either direction.

Then there are the young Christians encountering scientific cosmology for the first time in their education. I remember my own crisis when I first truly grasped the scale of the universe—not just intellectually but viscerally. Standing in a university lecture hall, seeing images from the Hubble Space Telescope, watching a visualization that zoomed out from Earth to solar system to galaxy to the cosmic web, I felt my childhood faith shrinking to a point and threatening to vanish. Many students experience this vertigo. Some abandon faith unnecessarily, thinking they must choose between the God of their youth and the universe of their textbooks. Others retreat into defensive fundamentalism, rejecting science to preserve faith. Both responses impoverish human understanding.

The third group consists of thoughtful believers who sense

that the universe revealed by science might enrich rather than threaten faith, but they don't know how to navigate the complexity. They're tired of simplistic harmonizations that ignore real tensions, but they're also unsatisfied with keeping science and faith in separate compartments. They intuit that truth must be unified, that the God who inspired the Psalms must be the same God who ignited the Big Bang, but they need help thinking through what this means.

The Approach of This Book

What I won't do in this book is pursue what scholars call concordism—the forced reading of modern science into ancient texts. Genesis isn't a physics textbook written in primitive language. The biblical authors weren't trying to answer our questions about quasars and quantum fields. They were addressing more fundamental concerns: Why does anything exist? What does it mean to be human? How should we live in relation to the Creator and creation? To make them speak to scientific questions is to disrespect both the texts and the questions.

Neither will I embrace the warfare model, that tired narrative that pits science against religion as eternal enemies. This story, largely manufactured in the nineteenth century for ideological purposes, doesn't reflect the actual history. Most of the founders of modern science were believers who saw their work as thinking God's thoughts after Him. Even today, plenty of excellent scientists maintain vibrant faith, and plenty of believers embrace scientific discovery. The conflict model serves those who profit from polarization, not those seeking truth.

But I also won't settle for easy compartmentalization, the comfortable fiction that science and faith occupy completely separate spheres that never interact. If Christianity is true, and if science discovers truth about nature, then they must somehow relate. The universe studied by physicists is the same creation proclaimed by faith. We can't permanently suspend the conversation between them without intellectual and spiritual schizophrenia.

Instead, I want to pursue what we might call critical dialogue—letting each domain speak in its own voice while seeking points of resonance and honestly acknowledging points of tension. Sometimes this dialogue produces harmony, like when Big Bang cosmology and creation ex nihilo seem to converge. Sometimes it generates productive friction, like when evolutionary suffering challenges simple notions of divine benevolence. Sometimes it leaves us with unresolved questions we must learn to carry, like how consciousness emerges from matter or how God acts in a law-governed universe.

What Modern Cosmology Actually Says

Before attempting any integration, we need to understand what contemporary cosmology actually claims, not what popularizers or apologists say it claims. The universe began 13.8 billion years ago in a hot, dense state—though "began" might be the wrong word since time itself seems to have started then. Space has been expanding ever since, not into something but as something, the fabric of reality itself stretching like bread dough rising.

The first atoms formed after 380,000 years when the universe cooled enough for electrons to settle into orbit around nuclei. Before this, the cosmos was opaque to light, a hot soup of particles and radiation. The first stars ignited after hundreds of millions of years, nuclear furnaces that forged heavy elements in their cores and scattered them through space when they died in supernovae explosions. These stellar ashes became the raw material for new stars, planets, and eventually, on at least one world, life.

Through processes we don't fully understand, chemistry became biology. Simple self-replicating molecules grew more complex. Single cells developed ways to work together. Life colonized every available niche, from deep ocean vents to mountain peaks. After billions of years of evolution—most of it microbial and invisible—consciousness emerged. Matter became aware of itself. The universe developed the capacity to question its own existence.

Perhaps most remarkably, we've discovered that everything we can see and touch—every star, planet, atom, and particle—constitutes less than 5% of what exists. The rest is dark matter, which holds galaxies together through gravity but doesn't interact with light, and dark energy, which drives the universe's accelerating expansion. We are, quite literally, cosmic minorities, made of the exceptional rather than the typical stuff of the universe.

What Christian Faith Affirms

Christianity makes claims that transcend physical description. The universe exists, Christians believe, because God freely chose to create it—not from necessity or loneliness but from generous love seeking others to love. Creation is pronounced good, meaningful, and purposeful, not the accidental byproduct of blind forces but the intentional artwork of divine creativity.

Humans, in this view, bear special significance as creatures made in God's image, capable of relationship with the divine. We're not just complex arrangements of atoms but persons with inherent dignity, moral responsibility, and eternal destiny. The Christian story insists that God doesn't merely observe creation from outside but enters it, most dramatically in the incarnation of Jesus Christ, where the infinite becomes finite, the eternal enters time, the Creator becomes creature.

Death, Christianity proclaims, is not final. The resurrection of

Christ previews creation's ultimate destiny—transformation into something that transcends current physical limitations while maintaining continuity with present existence. The universe itself will be renewed, liberated from bondage to decay, transformed into new creation where death no longer reigns.

These affirmations operate in different dimensions than scientific descriptions. Science addresses mechanism—how things work. Faith addresses meaning—why things matter. Science describes efficient causes—the physical processes that produce effects. Faith proclaims final causes—the purposes toward which things tend. Neither perspective is complete without the other, yet they don't easily combine.

Honest Difficulties We'll Face

This book won't pretend the integration is easy. We face serious challenges that resist simple resolution. The problem of scale haunts every attempt to maintain human significance: How can humans matter in a universe of two trillion galaxies? If we're the point, why such vast spaces empty of life? The Psalmist asked, "What is man that thou art mindful of him?" but he thought the universe was a cozy dome over a flat earth. The question becomes almost unbearable when we grasp cosmic immensity.

The problem of time is equally challenging. Why would God use 13.8 billion years of cosmic evolution to produce beings capable of relationship? Doesn't this seem wildly inefficient, even wasteful? Traditional theology spoke of God creating through divine fiat—"Let there be light," and immediately there was light. But we now know that the universe developed gradually, slowly, through processes that seem more like improvisation than implementation of a blueprint.

Then there's the problem of natural evil, which predates human sin by billions of years. Stars explode, destroying any planets in their vicinity. Galaxies collide with catastrophic consequences. On Earth, evolution proceeds through predation, parasitism, and mass extinctions. Suffering is built into the creative process. The same mechanisms that produced consciousness also produced cancer. How do we reconcile this with belief in a good Creator?

The problem of ultimate fate casts a shadow over all religious hope. Current physics predicts the universe will expand forever, growing colder and emptier until it reaches heat death—a state of maximum entropy where nothing interesting can ever happen again. If this is creation's destiny, what happens to Christian hope? How can we proclaim eternal life when physics predicts cosmic death?

Even fine-tuning, often cited as evidence for design, raises questions. Yes, the fundamental constants appear precisely calibrated for complexity and life. Change any of dozens of parameters slightly, and the universe becomes sterile. But does this point to a Designer, or does it merely reflect survivorship bias? Are we reading purpose into coincidence? And if God fine-tuned the universe for life, why is 99.99999% of it instantly lethal to life?

The Promise and the Risk

Despite these difficulties—or perhaps because of them—this exploration promises something valuable. We might develop a faith that engages reality as we actually find it, not as we wish it were. Christian faith has survived and even thrived through many cosmological revolutions. The early Christians had to reconcile their faith with Aristotelian cosmology. Medieval Christians integrated Ptolemaic astronomy. The Copernican revolution initially seemed to threaten Christianity but ultimately enriched it. Each transition was painful but productive.

The risk is that we might lose the particularity of Christian faith in cosmic generalities. In our eagerness to show that Christianity is compatible with modern cosmology, we might reduce the scandal of the incarnation to a cosmic principle or dilute the resurrection to a metaphor for emergent complexity. Christianity makes specific historical claims—that God became human in first-century Palestine, that Jesus physically rose from the dead, that these events changed the fundamental possibilities of existence. These claims can't be dissolved into scientific abstractions without losing their essence.

An Invitation to Wonder and Honesty

I write for those who refuse to choose between head and heart, between telescope and Scripture, between laboratory and sanctuary. If you've ever stood under the night sky feeling simultaneously insignificant and beloved, if you've ever wondered how prayer works in a universe governed by physical laws, if you've ever struggled to reconcile the God of your faith with the cosmos of your science classes, this book is for you.

We'll explore together, not as teacher and student but as fellow travelers puzzling over the same mysteries. I don't have all the answers—I'm not sure anyone does. But I've spent years living in the tension between scientific discovery and religious faith, and I've learned that the tension itself can be creative rather than destructive.

The cosmos revealed by modern science is stranger than our ancestors imagined—vaster, older, more complex, more beautiful, more terrible. Perhaps this means the God who created it (if God exists) is also greater than we've dared believe. Not the small, manageable deity of our comfort but the wild, untameable Mystery at the heart of existence. The God who spins galaxies and speaks in quantum uncertainties, who crafts consciousness from carbon and covenant from chaos.

Come with openness to surprise. Bring your questions and your doubts. We'll need intellectual honesty and spiritual courage for this journey. The path leads through difficult territory—cosmological vertigo, theological uncertainty, existential anxiety. But it might also lead to a faith that's stronger for being tested, a wonder that's deeper for being informed, a hope that persists not despite the universe science reveals but because of it.

The cosmos is calling. The Mystery awaits. Let's begin.

Revolution in Understanding the Universe

The End of the Eternal Cosmos

For most of human history, the universe appeared eternal and unchanging. The stars wheeled overhead in their fixed patterns, the same constellations that guided ancient mariners still guiding us today. Even Einstein, when his equations suggested an expanding cosmos, couldn't accept it. He added a cosmological constant to force his mathematics to describe a static universe, later calling it his greatest blunder. The discovery that the universe is expanding—and therefore had a beginning—represents one of the most profound paradigm shifts in human thought, comparable to realizing the Earth orbits the sun rather than vice versa.

I remember the moment this truly sank in for me. I was reading about Edwin Hubble's observations in the 1920s, how he noticed that galaxies are racing away from each other, and the farther away they are, the faster they're receding. The universe isn't just evolving; it's expanding. Space itself stretches, carrying galaxies apart like raisins in rising bread. And here's the thing that made me set down the book and stare at the wall: if you run the film backward, if you reverse the expansion, everything converges to a point of infinite density 13.8 billion years ago. The universe had a beginning. Time itself had a beginning.

For Christians, this discovery resonates with the doctrine of creation ex nihilo—from nothing. But we must be careful not to claim too much. The Big Bang describes the beginning of physical space and time as we know them, but it can't address the questions that keep philosophers and theologians awake at night: Why was there a singularity in the first place? Why did it expand? Why do physical laws govern its evolution instead of chaos? Why is there something rather than nothing? The convergence between Big Bang cosmology and theological creation is suggestive, even poetic, but it's not proof of anything. Science describes the "how" of the universe's beginning; faith addresses the "why."

The 95% We Don't Understand

Here's what should humble everyone, scientists and theologians alike: everything we can see—every star twinkling in the night sky, every planet, every galaxy revealed by our most powerful telescopes, every atom in our bodies—constitutes less than 5% of the universe. Let that sink in. We are minorities in our own cosmos, made of the exceptional stuff rather than the norm.

The rest? About 27% is dark matter, a form of existence we can detect only through its gravitational effects. It holds galaxies together—without it, they would fly apart like untethered merry-go-rounds. We know it's there with the same certainty that you know wind exists even though you can't see it, only its effects on leaves and flags. We can map dark matter's distribution, measure its gravitational lensing of light, see how it shapes the large-scale structure of the universe. But we don't know what it is. It's not atoms. It doesn't emit, absorb, or reflect light. It passes through ordinary matter like a ghost through walls, interacting only through gravity.

Dark energy is even stranger, if that's possible. It constitutes about 68% of the universe and drives the acceleration of cosmic expansion. Space itself seems to push against itself, and the more space expands, the more dark energy there is, causing ever-faster expansion in a kind of runaway feedback loop. We can measure its effects precisely—the 2011 Nobel Prize in Physics went to the teams that discovered the universe's accelerating expansion—but we have no idea what dark energy actually is. Some physicists think it's a property of space itself, that empty space isn't empty but seethes with quantum energy. Others propose more exotic explanations. The honest truth is we don't know.

This cosmic ignorance should breed humility in all of us. If we don't understand 95% of physical reality, how much less do we grasp ultimate reality? Scientists who confidently proclaim there's no need for God are standing on a tiny island of knowledge in an ocean of mystery. But religious believers who claim to know God's thoughts and plans should be equally humble. The universe is far stranger than either secular materialism or traditional theism imagined.

Fine-Tuning: Observation and Interpretation

The fundamental constants of nature appear precisely calibrated for the emergence of complexity and life. This isn't controversial—it's observational fact. The disagreement comes in interpreting what this means.

Consider the cosmological constant, fine-tuned to 120 decimal places. If it were slightly larger, the universe would have expanded too fast for galaxies to form. Slightly smaller, and the universe would have collapsed back on itself long ago. The strong nuclear force is calibrated to within 2%—any different and stars couldn't exist, or they'd burn out in moments rather than billions of years. The ratio of the electromagnetic force to gravity is precisely what's needed for stable stars that can nurture planets for eons. Carbon, the backbone of all known life, exists only because of a precise resonance in nuclear physics that allows stars to forge it from lighter elements. The initial entropy of the universe was tuned to one part in 10^10^123—a number so vast that if you wrote it out in full, the zeros would fill more volumes than there are atoms in the visible universe.

These are observations, not interpretations. The values are what they are. The question that divides people is: what, if anything, does this mean?

Some argue these precise values point to necessity—perhaps a deeper theory will show the constants couldn't be otherwise, that mathematical consistency requires exactly these values. But this pushes the question back a level: Why should reality be governed by that particular life-permitting theory rather than one of the infinite possible theories that wouldn't permit life?

Others invoke chance and the multiverse. In a vast or infinite cosmos with varying constants, some regions will randomly have life-permitting values. We naturally find ourselves in such a region because we couldn't exist elsewhere. This is logical but requires faith in cosmic vastness we can't observe. It's metaphysics dressed as physics, a belief system that avoids God by populating reality with infinite unobservable universes.

Still others see design, purpose, intention. The values are fine-tuned because they were chosen, like an engineer adjusting dials to achieve a desired outcome. This resonates with theism but isn't scientific proof—science can establish patterns but not purposes, correlations but not intentions.

I find myself drawn to the design interpretation, but I hold it lightly. Science has surprised us before. What seemed like design to one generation—the perfect circular orbits of planets, the ideal placement of Earth from the sun—found natural explanations in the next. Perhaps fine-tuning will too. Or perhaps it won't, and we're glimpsing something profound about the nature of reality, a hint that consciousness and purpose are built into the foundations rather than accidental byproducts.

The Universe That Doesn't Care

Let me be honest about something that troubles me: the cosmos revealed by modern astronomy often seems spectacularly indifferent to life, meaning, or purpose. Space is overwhelmingly empty—if Earth were a marble, the nearest star would be 200 miles away. Galaxies are islands in an ocean of nothing, and even within galaxies, stars are separated by vast voids. Most of the universe is hostile vacuum that would kill us instantly, our blood boiling, our lungs rupturing, our consciousness extinguished in seconds.

The universe is violent on scales that dwarf human comprehension. Stars explode with the force of billions of nuclear bombs detonating every second for weeks. Galaxies collide in slow-motion catastrophes lasting millions of years. Black holes devour everything that ventures too close, not even light escaping their grasp. Asteroids impact planets, gamma-ray bursts sterilize entire regions of space, supernovae flood their neighborhoods with lethal radiation. The universe created us through violence and will likely destroy us the same way.

And then there's the ultimate indifference: heat death. Stars will die, their nuclear fuel exhausted. Planets will be consumed by their expanding suns or flung into the frozen dark. Black holes will evaporate through Hawking radiation over timescales that make the current age of the universe seem like an eye-blink. The universe trends toward maximum entropy—eternal but empty, infinitely large but containing nothing of interest, a cosmic graveyard where nothing ever happens again.

At the quantum level, reality seems genuinely random rather than purposeful. Radioactive decay, quantum transitions, particle interactions—all probabilistic rather than determined. Einstein famously objected that God doesn't play dice with the universe, but experimental evidence suggests that at the fundamental level, dice are exactly what's being played. Where is divine providence in quantum randomness? Where is purpose in probabilistic wave functions?

These observations don't disprove God, but they challenge simple notions of cosmic purpose. The universe often seems more like a vast, beautiful, terrible accident than a carefully tended garden. If this is creation, it's not the cozy, human-centered creation our ancestors imagined. Christians must wrestle with this apparent indifference rather than explaining it away with easy theodicies or retreating into denial.

The Copernican Principle Extended

Copernicus removed Earth from the center of the cosmos, and we've been displaced ever further with each scientific advance. We're not at the spatial center—indeed, the universe has no center, every point equally valid as an origin of observation. We're not made of the universe's primary substance—dark matter and dark energy dominate while we're composed of rare baryonic matter. We're late arrivals, appearing in the last 0.0001% of cosmic history if you map the universe's age onto a single year. We're probably not unique—statistical arguments suggest life likely exists elsewhere among the hundreds of billions of planets in our galaxy alone. We're temporary—our species will go extinct, our planet will be destroyed, our sun will die, our galaxy will dissipate.

Each displacement has challenged human significance. Medieval cosmology placed Earth at the center (though they actually saw this as the lowest place, the bottom of the cosmic hierarchy with heaven at the periphery). Now we're peripheral in every sense—spatially, temporally, materially. We're cosmic afterthoughts, accidental byproducts, minor perturbations in fields that neither know nor care that we exist.

Yet I wonder if significance is really determined by centrality or size. A single conscious being capable of comprehending the cosmos might be more remarkable than billions of unconscious galaxies. After all, what is a galaxy but a collection of stars held together by gravity, following the laws of physics with no awareness, no choice, no appreciation of its own existence? But we—we fragile arrangements of carbon and water—we can contemplate infinity, create beauty, choose love over indifference, seek meaning in apparent meaninglessness. The universe becomes aware of itself through us. We are how the cosmos develops the capacity for wonder, gratitude, moral choice, and perhaps even worship.

Time, Change, and Divine Eternity

The discovery that the universe has a history—that it changes, develops, ages—raises profound theological questions. Classical theology, influenced by Greek philosophy, portrayed God as eternal and changeless, existing outside time in an eternal now where past, present, and future are equally present. But how does an unchanging God relate to a changing cosmos? How does the Eternal engage the temporal?

The Big Bang suggests time itself began, validating Augustine's ancient insight that God creates time along with the universe. There was no "before" the Big Bang because "before" is a temporal concept and time didn't exist yet. This is mind-bending but mathematically coherent. Yet it complicates everything about divine-human relationship. How do we pray to a God outside time? How does God respond to prayers that, from the eternal perspective, were always known? How does providence work if, from God's viewpoint, everything has already happened?

Moreover, the universe's expansion means new space continuously comes into existence. The cosmos is larger today than yesterday, will be larger tomorrow than today. Creation isn't just a past event but an ongoing process. Every moment, more universe exists than existed before. Does God create this new space? Is expansion itself a form of continuous creation? Are we witnessing Genesis not as ancient history but as present reality?

I don't have neat answers to these questions. The relationship between divine eternity and cosmic time remains mysterious, perhaps necessarily so. We're temporal beings trying to conceive the Eternal, finite minds grasping at the Infinite. It's like asking a character in a novel to understand the author who exists outside the story while simultaneously sustaining every word of it. We must hold both truths in tension: God transcends time yet acts in history, exists eternally yet engages the temporal, knows the end from the beginning yet somehow makes room for genuine relationship and real choice.

Multiple Cosmologies, Multiple Challenges

Contemporary physics entertains various cosmological models, each with different implications for faith. We don't yet know which, if any, is correct, and each poses unique challenges to traditional theology.

Eternal inflation suggests that quantum fluctuations continuously spawn new universes, our Big Bang just one of countless others, creation an endless process rather than singular event. This challenges the uniqueness of creation and raises questions about the scope of salvation history. Is our universe special, or just one random bubble in infinite foam?

Cyclic cosmology proposes the universe undergoes endless cycles of expansion and contraction, Big Bang followed by Big Crunch followed by Big Bang again, forever. This eliminates the need for an absolute beginning and challenges linear salvation history. If the universe repeats eternally, where is the hope for transformation and new creation?

The multiverse hypothesis suggests multiple universes exist with different physical laws, different constants, different possibilities. Fine-tuning becomes a selection effect rather than evidence of design. Incarnation becomes radically particular—why this universe rather than others? The scope of redemption becomes problematic—does salvation extend across all universes or only ours?

Some physicists seriously discuss the simulation hypothesis—that our universe might be a simulation running in some deeper reality, physical laws just programming code, existence itself virtual. This poses obvious challenges to doctrines of creation, incarnation, and redemption. If we're simulations, what does it mean to be made in God's image? Can simulated beings have souls? Did Christ become simulated flesh?

Christians shouldn't panic over these possibilities. They're speculative extrapolations, not established facts. Most are untestable and might remain forever in the realm of mathematical possibility rather than physical reality. But they remind us that our current cosmological understanding may be provisional. Faith must be robust enough to engage whatever cosmology emerges, not tied to particular scientific theories that might be overturned.

Wonder Despite Difficulty

Despite these challenges—or perhaps because of them—the universe evokes profound wonder. Dark matter reminds us that most of reality is hidden from our senses, that the visible is exception rather than rule. Dark energy shows creation is still happening, space itself expanding, the universe growing larger every moment. Fine-tuning hints at deeper order, mathematical elegance that seems to transcend necessity. Even heat death testifies to time's preciousness, making each moment valuable precisely because moments won't last forever.

This wonder doesn't prove God but it opens space for faith. A universe that exceeds our understanding, that surprises us with beauty and terror, that generates consciousness from quantum fields—such a universe is at least consonant with belief in transcendent Mystery. The cosmos doesn't compel belief but it permits it, even invites it.

The revolution in cosmological understanding doesn't provide easy answers for faith. It complicates simple narratives, challenges traditional assumptions, raises new questions. But it also reveals a cosmos far grander, stranger, and more wonderful than our ancestors imagined. If God exists, this God must be greater than we thought—not a cosmic engineer tinkering with dials but the Source of existence itself, the Mystery within mystery, the One in whom quantum fields and galaxies alike live and move and have their being.

Living with Cosmic Vertigo

Modern cosmology induces a kind of vertigo. The scales are too vast, the implications too profound, the uncertainties too deep. We thought we stood on solid ground and discovered we're spinning through space on a tiny rock, made of minority matter, in an expanding universe we barely understand. This vertigo is appropriate—it reflects our actual situation.

We are cosmically small, temporally brief, and fundamentally ignorant about most of reality. Any faith that doesn't acknowledge this is self-deception. Yet within this vertigo, we find ourselves capable of something remarkable: understanding our situation. We can measure the universe's age, map its structure, trace its history, and even predict its future. We can ask why it exists and what it means. We can choose how to respond—with despair or wonder, nihilism or faith.

The revolution in cosmological understanding doesn't resolve the question of God. It reframes it. The God who might exist must be compatible with quantum mechanics and general relativity, with dark energy and heat death, with 13.8 billion years of cosmic evolution and two trillion galaxies. This isn't the small God of easy answers but the vast God of ultimate mystery—perhaps more worthy of worship than we ever imagined, perhaps more terrifying too.

Standing under the night sky, aware now of what those points of light really are—nuclear furnaces millions of times larger than Earth, separated by distances that light takes years to cross, arranged in galaxies racing apart in an expanding cosmos heading toward heat death—I feel simultaneously smaller and more significant than ever. Smaller because the universe dwarfs human scales beyond comprehension. More significant because somehow this vast, impersonal cosmos produced beings capable of personal relationship, moral choice, and spiritual longing.

This is the paradox we must live with: we are nothing and everything, accidental and intended, insignificant specks and the universe's way of understanding itself. The cosmos doesn't resolve this paradox but intensifies it. And perhaps that's exactly what we should expect from a universe strange enough to exist at all.

Spacetime and Divine Eternity

The Fabric of Reality Itself

Einstein's discovery that space and time are interwoven into spacetime, that this fabric can bend and stretch, that time flows at different rates in different places—this isn't just a revision of Newton. It's a fundamental reconceptualization of reality's structure. When I first truly understood this, not just as equations on a page but as a description of the world I inhabit, I had to sit down. The solid ground beneath my feet suddenly felt less solid, the steady tick of my watch less steady.

Time doesn't tick uniformly throughout the universe. If you stand at the base of a mountain, time passes slightly slower for you than for someone at the peak, because you're deeper in Earth's gravitational field. This isn't science fiction or theoretical speculation—it's measurable reality. GPS satellites must account for time dilation or they accumulate navigational errors of several miles per day. The satellites' clocks run faster than Earth-bound clocks by about 38 microseconds daily. Without

Einstein's corrections, GPS would be useless within minutes.

What does this mean for prayer? When I pray, am I speaking to a God who experiences time as I do, moment by moment? Or to a God who sees all moments simultaneously, like viewing a landscape from above rather than walking through it? The traditional answer—that God is eternal and sees all time at once—suddenly becomes more plausible but also more problematic. If God already knows not just what I will pray but has already, from the eternal perspective, answered or not answered, what is the point of praying at all?

Augustine's Lucky Guess?

Sixteen centuries before Einstein, Augustine of Hippo argued that time itself is part of creation. Skeptics had asked what God was doing before creating the world—was the Almighty just sitting around for infinite eons before deciding to create? Augustine's response was philosophically sophisticated: there was no "before" creation because time came into existence with the universe. God doesn't exist in time that predates creation but creates time along with everything else.

This wasn't based on physics or mathematics but on theological necessity. If God created everything, and time is something rather than nothing, then God must have created time. Augustine reasoned his way to a conclusion that modern cosmology would confirm: time had a beginning, and asking what happened "before" time is like asking what's north of the North Pole—the question contains a category error.

Was Augustine lucky? Divinely inspired? Or is there something about careful theological reasoning that can reach truths that science later confirms? I'm wary of drawing strong conclusions from this single convergence. Theology has been wrong about plenty of things that science later corrected. But it's intriguing that thinking seriously about creation and eternity led Augustine to insights about time that anticipated Einstein by more than a millennium.

The Block Universe: A Theological Problem?

If relativity is correct—and every experiment confirms it—then past, present, and future all exist simultaneously in four-dimensional spacetime. This is called the "block universe" or "eternalism." From this perspective, time doesn't "flow" from past to future. Instead, all moments exist eternally, and our experience of temporal flow reflects our embedded perspective as we trace our worldline through spacetime. We're like trains on tracks, experiencing the landscape sequentially even though the entire landscape exists simultaneously.

This seems to support classical theism's view of divine eternity. God sees all moments at once, like viewing a parade from above rather than from the street. My birth, life, and death are equally present to God's eternal awareness. From the divine perspective, Abraham, Jesus, and whoever will be alive in the year 3000 all exist simultaneously.

But this creates serious problems that keep me awake at night. If the future already exists in the block universe, are our choices real or illusory? When I stand at a crossroads and choose left rather than right, am I really choosing, or am I simply discovering what I was always going to choose? The block universe seems deterministic—our future choices already "exist" in spacetime, waiting for us to arrive at them.

What's the point of petitionary prayer in a block universe? When I pray for healing, for guidance, for intervention, I'm asking God to change what, from the eternal perspective, already exists. It's like asking an author to change chapter 10 when the whole book is already written. Prayer seems to become cosmic theater, playing out a script that's already fixed.

And how can God have genuine relationship with creatures

if, from the divine perspective, everything has already happened? Relationship requires real interaction, response, development. But in the block universe, there's no development, only the illusion of development as we trace our predetermined paths through spacetime.

The problem of evil becomes even more acute. If all moments are equally real to God, does suffering ever end? From the eternal perspective, is the Holocaust always happening, every moment of agony eternally present to divine awareness? The child dying of cancer—does that suffering exist forever in the block universe, never truly past, never truly healed?

When Physics and Faith Diverge

Let me be honest about where modern physics seems to conflict irreconcilably with Christian convictions. These aren't minor puzzles but fundamental tensions that I haven't resolved despite years of thinking about them.

Divine action is perhaps the most pressing problem. If spacetime is a closed system governed by mathematical laws, where and how does God act? Some suggest quantum indeterminacy provides openness for divine action, but this seems both too subtle and too random. A God who works only through quantum uncertainties is hardly the God who parted the Red Sea or raised Jesus from the dead. And influencing quantum events to produce specific macroscopic outcomes would require such massive intervention that it would be detectable—yet we detect no such influence.

Christianity is irreducibly historical. Creation, fall, incarnation, crucifixion, resurrection, eschaton—these aren't timeless truths but events that happened (or will happen) at particular moments. They require genuine temporal succession, real before and after, actual development and change. But the block universe suggests all this succession is illusory, that from the proper perspective all these "events" exist simultaneously in the eternal now.

This saps moral urgency from our choices. If all moments exist eternally, why does it matter what I choose? My good and bad actions are equally permanent fixtures of spacetime. The murder and the mercy both exist forever. Repentance can't change what eternally is. Justice and injustice are frozen in the ice of spacetime, unchangeable, unredeemable.

Christian hope involves genuine newness—new creation, new heaven and new earth, transformation of what is into what should be. The block universe seems to preclude real novelty. Everything that will be already is. The future exists as definitely as the past. Where is room for surprise, for grace, for redemption that actually changes things rather than merely revealing what always was?

Multiple Times, Multiple Problems

Relativity reveals not one universal time but multiple times, each valid in its own frame of reference. There's proper time—what each observer experiences along their worldline through spacetime. There's coordinate time—the arbitrary labeling we use for calculations. There's cosmic time—the universe's "age" as measured from the Big Bang. There's thermodynamic time—the direction of entropy increase. There's psychological time—our subjective experience of duration, how minutes crawl during boredom and fly during joy.

Which time does God inhabit? All of them? None? Some divine time we can't conceive? The multiplication of times complicates simple notions of divine eternity "outside" time. Perhaps God relates to different times differently—transcending cosmic time while entering proper time in the incarnation, experiencing psychological time in divine empathy with creatures.

But this multiplication of relationships seems to fragment divine unity. Are there multiple divine perspectives corresponding to multiple times? Does God experience my proper time with me while simultaneously transcending it? The mathematics of relativity are clear, but the theology becomes murky.

Quantum Complications

Just when relativity seems to establish a deterministic block universe, quantum mechanics introduces fundamental indeterminacy. The future might not exist definitely but in superposition until measurement collapses possibilities into actualities. Schrödinger's cat is both alive and dead until we open the box. Perhaps the future is similarly undetermined until we "measure" it by arriving there.

But this creates its own problems. How do we reconcile relativity's block universe with quantum indeterminacy? Different interpretations of quantum mechanics have different implications. The Copenhagen interpretation preserves openness but makes reality fundamentally fuzzy. Many Worlds suggests all possibilities exist in parallel universes, which multiplies realities beyond comprehension. Hidden variable theories restore determinism at the cost of non-locality. Relational interpretations make consciousness fundamental to reality in ways that seem to privilege observers.

We don't know which interpretation is correct. Building theological arguments on any particular interpretation is like building on sand—the next experimental result might wash it away. Yet we can't simply ignore quantum mechanics while thinking about time, eternity, and divine action. The universe is quantum mechanical at its base, and any theology that ignores this is incomplete.

The Incarnation: When Eternity Enters Time

The incarnation represents the ultimate intersection of eternity and time. Christian faith proclaims that in Christ, the eternal Word became temporal flesh, infinite God became finite human, the Creator entered creation. From a physics perspective, this seems impossible. How can an eternal being experience temporal succession? How can infinity be contained in finite spacetime? How can the sustainer of quantum fields become a configuration of quantum fields?

Yet perhaps the incarnation points toward possibilities physics hasn't imagined. Maybe the eternal and temporal aren't mutually exclusive but can interpenetrate in ways our mathematics can't capture. Maybe consciousness transcends the categories of classical spacetime. Maybe the incarnation reveals something about reality's structure that neither physics nor theology alone can grasp.

When I think about the incarnation now, informed by relativity and quantum mechanics, it seems both more impossible and more necessary than ever. More impossible because we know how intricately time is woven into the fabric of reality—to be human is to be temporal, experiencing duration, change, growth, and decay. More necessary because if God wants genuine relationship with temporal creatures, divinity must somehow enter our time-bound existence.

The incarnation is either the deepest truth or the greatest absurdity. God walking the dusty roads of Palestine while simultaneously sustaining the universe. The Word through whom galaxies spin becoming a baby who needs milk and sleep. Infinite consciousness constrained to finite neural networks. It makes no sense. And yet... and yet... something in me recognizes truth that transcends sense.

Prayer Across Spacetime

If God transcends spacetime, prayer doesn't need to traverse distance or duration. We're not sending messages across cosmic distances like radio signals that take time to arrive. We're already in God's presence—indeed, there's no "where" where God is not present. Prayer is not about informing God or traveling to God but about becoming aware of the divine presence that already pervades every point of spacetime.

But this raises its own questions. If God eternally knows our prayers, why pray? If the future exists in the block universe, what can prayer change? How can temporal beings genuinely communicate with an eternal being? It's like characters in a book trying to talk to the author—the author knows every word they'll say because the author wrote those words, yet somehow the characters need to say them.

I've come to think of prayer less as communication and more as participation. Not sending messages to distant God but aligning myself with divine purpose that pervades reality. Not informing God of my needs but becoming conscious of God's consciousness of those needs. Not changing God's mind but letting God's mind change mine. This isn't fully satisfying—part of me still wants prayer to be genuine dialogue, real interaction, actual influence. But in the strange world of spacetime physics, participation might be the most coherent model available.

Living in the Tension

Rather than forcing resolution, I'm learning to live creatively with the tension between physics and theology regarding time. Both domains are grasping at something beyond full human comprehension. Physics describes patterns and relationships but can't explain why there's something rather than nothing, why these laws rather than others, why consciousness emerges from matter. Theology offers meaning and purpose but struggles to explain how eternal God relates to temporal creation, how divine action works in law-governed nature, how prayer functions in determined or quantum-uncertain reality.

Perhaps the tension itself is telling us something. Maybe reality is too rich, too complex, too strange for any single framework to capture. Light is both wave and particle—not sometimes one, sometimes the other, but always both simultaneously. Maybe divine action is both eternal and temporal, both transcendent and immanent, both determining and liberating, in ways that exceed our either-or categories.

When I pray now, I hold multiple perspectives simultaneously. I speak as if the future is open and my prayers matter, because from my embedded perspective in spacetime, they do. But I also rest in the trust that from God's eternal perspective, my prayers are already woven into the fabric of reality, already part of how God sustains and guides creation. This isn't fully coherent, but then neither is the universe.

Eschatological Time

Christian hope involves transformation—new creation, resurrection, eternal life. How does this relate to spacetime physics? Current physics describes spacetime as heading toward heat death, maximum entropy, a state where nothing interesting ever happens again. This seems to contradict Christian eschatology.

But the resurrection offers a paradigm. Christ's risen body transcended normal spacetime constraints while remaining physical. He appeared and disappeared, passed through walls, traveled instantly, yet could be touched and could eat. Perhaps new creation involves similar transformation—not abandonment of spacetime but its fulfillment in ways physics can't anticipate.

This requires enormous faith. We're essentially betting that God can do something physics says is impossible—reverse entropy, transform spacetime itself, create genuinely new possibilities. But then, existence itself is impossible. Why is there something rather than nothing? Why these laws? Why consciousness? If the impossible already happened in creation, perhaps it can happen again in new creation.

Time, Eternity, and Humility

Modern physics reveals time to be far stranger than common sense suggests—relative, flexible, possibly emergent rather than fundamental, interwoven with space into a four-dimensional block that might or might not be deterministic. This complicates theological discussions about divine eternity, providence, prayer, and eschatology.

I could force compatibility by reinterpreting theology to fit physics or dismissing physics where it challenges theology. But intellectual honesty requires admitting the tensions and living with unresolved questions. We're temporal beings trying to understand eternity, finite minds grasping at infinity, creatures of spacetime attempting to comprehend what transcends spacetime.

The gap between human understanding and ultimate reality is probably greater than either physics or theology alone acknowledges. We see through a glass darkly, as Paul wrote, and the glass is darker than we thought. Both our physics and our theology are provisional, incomplete, pointing toward truths we can't fully grasp.

Yet we continue both scientific investigation and religious practice, not because we've resolved all tensions but because both are essential to human flourishing. Science feeds our curiosity about how reality works. Faith nourishes our need for meaning, relationship, and hope. The conversation between them is difficult, sometimes painful, but necessary.

The universe has temporal structure more complex than our ancestors imagined. If God exists, this God must relate to time in ways that transcend our simple categories—neither simply "in" time nor simply "outside" it but somehow both and neither and more. Perhaps that's precisely what we should expect of the Mystery at the heart of reality. A God simple enough for us to fully understand wouldn't be God at all.

Quantum Reality and Divine Action

The Death of Determinism

Classical physics portrayed a clockwork universe. Laplace famously claimed that an intelligence knowing the position and momentum of every particle could predict the entire future and retrodict the entire past. The universe was a vast machine, grinding forward according to inexorable laws. This left little room for genuine novelty, freedom, or divine action that didn't violate natural laws. If you wanted God to act, you needed God to break the rules, to intervene miraculously in the otherwise closed system.

Quantum mechanics shattered this deterministic picture, but what it offers instead remains deeply mysterious and contested. At the quantum level, events are probabilistic rather than determined. We can calculate the likelihood of finding an electron in a particular location, but we cannot predict with certainty where it will be. When I measure the spin of an electron, I can know it has a 50% chance of being up and 50%

chance of being down, but nothing in the universe determines which it will be until I measure it. A radioactive atom might decay in the next second or the next million years—the timing is genuinely random within statistical patterns.

Does this quantum indeterminacy create "space" for divine action or free will? I used to think so, but I've become more cautious. Random is not the same as free. If my decisions were based on quantum coin flips in my neurons, they would be arbitrary, not free in any meaningful sense. A choice driven by quantum randomness is no more "mine" than one driven by classical determinism. And the idea of God acting through quantum events faces its own problems—divine action reduced to manipulating probability distributions seems to diminish God to a hidden variable in equations, a cosmic thumb on quantum scales.

The Measurement Problem and Its Mysteries

The central mystery of quantum mechanics—the one that kept Einstein awake at night and still troubles physicists today—is measurement. Before observation, quantum systems exist in superposition, multiple states simultaneously. An electron passing through a double slit goes through both slits at once, interfering with itself. Schrödinger's infamous cat is both alive and dead until we open the box. But when we observe, when we measure, the superposition "collapses" into a definite state. The electron is detected at a particular location. The cat is either alive or dead, not both.

But what counts as observation? Does it require consciousness? Can an unconscious measuring device collapse a wave function, or does it take a conscious observer to make reality definite? How does nature "decide" which possibility becomes actual? After nearly a century of quantum mechanics, we still don't have agreed-upon answers.

The various interpretations have profoundly different implications for theology. The Copenhagen interpretation, still the most common, says reality is fundamentally indeterminate until measured. This preserves openness and possibility but makes reality oddly dependent on observation. Are we co-creating reality through our observations? Is consciousness more fundamental than matter?

The Many Worlds interpretation eliminates the measurement problem by saying that all possibilities occur in parallel universes. When I measure the electron's spin, the universe branches—in one branch I observe spin-up, in another branch another version of me observes spin-down. This eliminates randomness but at the cost of infinite proliferating realities. Every quantum event spawns new universes. The implications for theology are staggering. Are there infinite versions of me, some saved, some damned? Did Christ die in all branches or only some?

Hidden variable theories suggest underlying deterministic factors we can't access govern quantum events. This restores determinism but requires accepting invisible influences that seem to violate relativity's prohibition on faster-than-light effects. If God is the hidden variable, we're back to divine determinism, just hidden at a deeper level.

Relational interpretations suggest properties exist only relative to observers. Reality is fundamentally relational rather than objective. This resonates with theological emphasis on relationship but challenges the notion of objective truth. Is reality itself perspectival? Does God have a perspective, or does divine observation create the only truly objective reality?

The Quantum-Spiritual Temptation

The mysteriousness of quantum mechanics tempts religious thinkers to make connections that aren't warranted. I've read too many books claiming quantum mechanics proves consciousness creates reality, that we can manifest our desires through quantum intention, that entanglement explains the efficacy of prayer. This is almost always nonsense, a misunderstanding of both quantum mechanics and spirituality.

"Observation" in quantum mechanics doesn't mean conscious observation—it means any physical interaction that transfers information. An electron "observes" another electron when they interact. A photon "measures" an atom when it's absorbed. Consciousness isn't required. The universe was collapsing wave functions for billions of years before conscious observers evolved.

The idea of quantum healing—that consciousness can direct quantum events to heal bodies—misunderstands both quantum mechanics and biology. Quantum effects typically "decohere" in warm, wet, noisy biological systems. By the time you get to anything as large as a cell, quantum superposition has long since collapsed into definite states. The brain is far too warm and noisy to maintain quantum coherence. If quantum effects influence neural activity, it's through randomness, not through conscious control.

Quantum entanglement is real and verified—particles can be correlated in ways that seem to transcend space. When I measure one particle's spin, I instantly know its entangled partner's spin, regardless of distance. Einstein called this "spooky action at a distance" and it troubled him deeply. But entanglement doesn't mean everything is connected to everything else in some cosmic web of consciousness. Entanglement is fragile, easily broken by interaction with the environment. It doesn't transmit usable information. It can't explain prayer or psychic phenomena or universal consciousness.

Entanglement: Real but Limited

Let me be clear about what entanglement actually means, because it's so often misunderstood. When two particles are entangled, measuring one instantly affects the other, regardless of distance. This is experimentally verified and technologically useful—quantum computers depend on it. But the effect is subtle and limited.

Entanglement does show that reality is non-local in ways classical physics never imagined. The universe isn't just a collection of independent objects interacting through forces. There are irreducible correlations, connections that transcend space. This might resonate with theological themes of interconnection and unity—we are members of one body, Paul wrote, and quantum mechanics hints that this might be more than metaphor.

But entanglement doesn't allow faster-than-light communication. I can't send a message through entangled particles. It doesn't mean everything is connected to everything—most things aren't entangled with most other things. It doesn't prove universal consciousness or explain the communion of saints or validate New Age spirituality. The lesson of entanglement is more modest but still profound: reality has dimensions of connection our everyday experience doesn't reveal. The universe is more unified than classical physics suggested, though not in the mystical way some claim.

Divine Action: Beyond the Gaps

How might God act in a quantum world without violating natural laws? This question has occupied theologians and religious physicists for decades, producing various proposals, each with significant problems.

Some suggest God determines which quantum possibility

becomes actual while respecting probability distributions. God doesn't violate quantum mechanics but works within its probabilistic structure, like a pianist improvising within the constraints of the instrument. But this makes God a hidden variable, and most hidden variable theories have been experimentally ruled out by tests of Bell's inequalities. If God is determining quantum outcomes, the divine action is hidden in ways that seem to require conspiracy-level fine-tuning.

Others propose that God acts by inputting information without energy, similar to how measurement updates wave functions without adding energy to the system. But information transfer typically requires energy—that's basic thermodynamics. And this model still seems interventionist, God reaching into the system from outside rather than working through natural processes.

The idea of top-down causation suggests God influences quantum events through higher-level organizing principles, the way consciousness seems to influence neural activity without violating physics at the molecular level. But this is more assertion than explanation. We don't understand how consciousness relates to neural activity, so using it as a model for divine action is building mystery upon mystery.

Perhaps we're approaching this wrong. Instead of looking for gaps where God might sneak action into nature, we should recognize that if God exists, divine action must be compatible with what we observe—including quantum mechanics. The question isn't "How does God violate quantum mechanics?" but "How does quantum mechanics express divine action?" Not "Where are the gaps?" but "How is God present in natural processes?"

The Quantum Vacuum: Something from "Nothing"?

One of the most theologically suggestive discoveries of quantum field theory is that empty space isn't empty. The quantum vacuum seethes with virtual particles appearing and disappearing, borrowing energy from the uncertainty principle for times too brief to measure directly. This zero-point energy of the vacuum has real effects—it produces the Casimir force between parallel plates, contributes to the cosmological constant, and may have driven cosmic inflation.

Some physicists and theologians have suggested this shows how something can come from nothing, relevant to creation ex nihilo. But the quantum vacuum isn't "nothing" in the philosophical sense. It's the ground state of quantum fields, existing in spacetime, governed by physical laws. It's very much something, just not something made of permanent particles.

The quantum vacuum might illustrate creation's continuing nature—reality constantly emerging from potentiality, every moment a kind of mini-creation as virtual particles pop into existence. But it doesn't explain why there are quantum fields in the first place, why physical laws govern them, or why anything exists at all. The theological question of creation ex nihilo—why something rather than nothing—operates at a different level than physics.

Quantum Mechanics and Natural Theology

Despite the difficulties and temptations to overreach, quantum mechanics does contribute to natural theology—learning about God through nature—in legitimate ways.

Reality exceeds common sense in ways that should make us humble about pronouncing what's possible or impossible. Before quantum mechanics, who would have believed that particles could be in two places at once, that observing one particle could instantly affect another across the universe, that reality could be fundamentally probabilistic rather than determined? Quantum mechanics shows that reality is stranger than everyday experience suggests. This openness to mystery, to paradox, to the seemingly impossible, resonates with theological recognition of divine transcendence.

Quantum mechanics suggests relationship is more fundamental than substance. Properties emerge from relationships rather than existing independently. An electron doesn't have definite position or momentum until it interacts with something else. Reality seems to be fundamentally relational, which echoes theological emphasis on relationship—God as Trinity, humans as imago Dei through relationship, love as the fundamental force. Though we shouldn't push the parallel too far, there's something suggestive about a universe that's relational all the way down.

The complementarity principle in quantum mechanics—that mutually exclusive descriptions like wave and particle are both needed for complete understanding—might illuminate theological paradoxes. Just as light is both wave and particle, perhaps divine action is both transcendent and immanent, both eternal and temporal. Not sometimes one, sometimes the other, but always both simultaneously in ways that transcend our either-or logic.

Living with Quantum Uncertainty

What practical difference does quantum mechanics make for Christian faith? After years of studying both physics and theology, I've come to several conclusions.

First, humility about reality. If quantum mechanics is correct—and every experiment confirms it—then reality is fundamentally stranger than we imagined. Our theological formulations, based on common-sense categories from everyday experience, might need radical revision. God might act in ways that transcend not just our knowledge but our categories of knowledge.

Second, comfort with mystery. Quantum mechanics shows that mystery isn't failure of understanding but might be built into reality's structure. We can predict quantum behavior with extraordinary precision while having no idea what's "really" going on. Similarly, we might experience God's action without understanding its mechanism. Mystery isn't a problem to be solved but a reality to be acknowledged.

Third, caution about claims. We should resist using quantum mechanics to "prove" spiritual claims. The connections are usually more poetic than logical, more suggestive than demonstrative. Quantum mechanics makes space for theological concepts but doesn't establish them. The universe is open to divine action but doesn't prove it occurs.

Fourth, wonder at creation. The quantum realm reveals extraordinary elegance and surprise. That conscious beings emerged from quantum fields through billions of years of evolution is cause for profound wonder, whether you're religious or not. The equations of quantum mechanics are beautiful in ways that seem excessive if the universe is purely functional. Why should mathematics developed by human minds describe reality so perfectly? Why should reality be comprehensible at all?

The Observer and the Observed

The role of observation in quantum mechanics raises philosophical questions that overlap with theology. Does observation partly create reality? Is consciousness fundamental rather than emergent? How do observer and observed relate?

These questions resonate with theological themes. Humans as

creation's consciousness, the universe becoming aware of itself through us. The importance of attention and awareness in spiritual practice. The participatory nature of knowledge—we don't just discover truth but participate in its disclosure. The relationship between divine observation and creation's existence—does God's observing sustain reality?

But we must be careful not to make quick equations. The "observer" in quantum mechanics need not be conscious—any interaction that transfers information counts as observation. And divine "observation" is categorically different from physical measurement. God doesn't collapse wave functions the way physicists do.

Still, quantum mechanics suggests that the sharp distinction between observer and observed, subject and object, knower and known, might be an approximation rather than fundamental. Reality might be more participatory than we thought. We're not outside the universe looking in but part of it, and our participation partly constitutes what it is. This opens conceptual space for prayer, providence, and divine-human cooperation, even if it doesn't explain them.

Quantum Caution and Wonder

Quantum mechanics reveals a reality that transcends common-sense categories. Particles that are waves, events without definite causes, connections that transcend space, observation that partly creates what's observed—the quantum world is deeply strange. This should make us both cautious and wonder-filled.

Cautious because quantum mechanics is often misused to support dubious spiritual claims. The desire to find God in quantum gaps, to prove consciousness is fundamental, to validate religious experience through physics—these temptations are strong but should be resisted. The connections between quantum and spiritual are usually more poetic than logical. We must avoid the intellectual sleight-of-hand that makes mystery in one domain seem to explain mystery in another.

Wonder-filled because the quantum realm reveals nature's extraordinary subtlety and beauty. That consciousness emerges from quantum fields, that atoms obey probability laws that somehow create stable matter, that measurement transforms possibility into actuality—these are profound mysteries that evoke appropriate awe. The universe is far stranger and more wonderful than our ancestors imagined.

Quantum mechanics neither proves nor disproves God. It doesn't explain consciousness, free will, or divine action. But it does show that reality has depths our everyday experience doesn't reveal, that the universe is open rather than closed, relational rather than mechanical, mysterious rather than fully comprehensible.

In the quantum realm, we all stand before mystery. The atheist physicist and the believing physicist face the same paradoxes, the same interpretive choices, the same limits of understanding. And in that shared unknowing, that common wonder at reality's strangeness, there might be space for both scientific investigation and religious faith to continue their distinct but related quests for understanding.

The universe is quantum mechanical at its base, and any theology that ignores this is incomplete. But quantum mechanics is also incomplete, unable to explain its own foundations, its relationship to consciousness, its place in larger reality. Both science and faith point beyond themselves to truths neither fully grasps. And perhaps that's as it should be. A reality simple enough for us to fully understand wouldn't be worth understanding.

Different Paths of Integration

The Plurality of Approaches

There is no single correct way to relate cosmology and Christian faith. This used to frustrate me—I wanted the answer, the solution, the way to hold these truths together without contradiction. But I've come to see the plurality of approaches not as failure but as recognition of complexity. Different people, facing different questions, living in different contexts, need different ways of navigating the tension between scientific discovery and religious conviction.

Walking through a university campus, you'll find physicists who are devout believers and others who are committed atheists, working side by side on the same experiments, publishing in the same journals. In any given church, you'll find some members who embrace evolution and others who reject it, some who see fine-tuning as evidence of God and others who see it as coincidence, all singing the same hymns, sharing the same communion. The diversity of approaches reflects the diversity of human experience and the complexity of the questions we're asking.

This chapter examines major integration strategies, not to declare a winner but to map the territory. I want to present each approach sympathetically, understanding why thoughtful people choose it, while also acknowledging its limitations. Perhaps you'll recognize your own approach here, or discover one that resonates more deeply, or realize you've been combining elements from several without knowing it.

Independence: Non-Overlapping Magisteria

Stephen Jay Gould, the late paleontologist and science popularizer, proposed that science and religion occupy "Non-Overlapping Magisteria" (NOMA). Science addresses questions of fact—how the universe works, what it's made of, how it developed. Religion addresses questions of meaning and morality—why we exist, how we should live, what ultimately matters. Neither should trespass on the other's territory. Science can't tell us whether God exists because God isn't a scientific hypothesis. Religion can't dictate scientific conclusions because revelation doesn't address physical mechanisms.

I know a physicist who lives this separation with remarkable consistency. In her laboratory, she studies quantum chromodynamics, working out the equations that govern quarks and gluons inside atomic nuclei. She never thinks about God while calculating cross-sections or analyzing collision data. On Sunday mornings, she worships at a Presbyterian church, sings in the choir, and teaches adult education classes on prayer and spiritual practice. She never mentions quantum mechanics in church.

When I asked her how she holds these worlds together, she smiled and said she doesn't hold them together—that's the point. "They're asking different questions," she explained.

"Physics asks how. Faith asks why. Physics describes mechanisms. Faith provides meaning. I don't need them to connect because they're not trying to occupy the same space."

The strength of this approach is its clarity. It prevents conflicts by maintaining boundaries. Science can pursue its methods without theological interference. Faith can maintain its convictions without scientific challenge. Each discipline preserves its integrity. For working scientists who are believers, this approach offers professional credibility—colleagues can't accuse them of letting religion influence their research. For believers who accept mainstream science, it offers theological peace—they don't have to reconcile every scientific discovery with Scripture.

But complete separation seems artificial to me. If God created the universe, shouldn't creation somehow reflect divine character? If humans are purely physical beings, doesn't neuroscience affect our understanding of soul, consciousness, and moral responsibility? The boundaries blur when we examine them closely. Cosmology raises questions about ultimate origins that verge on theology. Theology makes claims about reality that seem to have empirical implications. Can we really keep them entirely separate?

Dialogue: Consonance Without Concordance

This approach seeks meaningful conversation between science and faith without forcing agreement. Each discipline maintains its own methods and authority while looking for resonances, mutual illumination, and creative interaction. It's like musicians improvising together—each plays their own instrument, follows their own musical logic, but listens for harmonies and responds to what the other is playing.

I think of a theologian friend who regularly audits physics courses at the local university. She doesn't pretend to master the mathematics, but she wants to understand what physicists actually claim about reality. When she teaches systematic theology, she incorporates scientific insights not as proof texts but as conversation partners. The Big Bang doesn't prove creation ex nihilo but it does resonate with it, and exploring that resonance deepens both scientific and theological understanding.

Similarly, I know scientists whose faith inspires their research questions without determining their answers. A cosmologist wonders at fine-tuning because of his religious convictions, but he investigates it with the same rigor he'd apply to any phenomenon. An evolutionary biologist sees her work as exploring God's creativity, but she follows the evidence wherever it leads, even when it challenges traditional religious narratives.

The dialogue approach respects both scientific and theological integrity. It allows for genuine learning in both directions—theology informed by scientific discovery, science inspired by theological imagination. It acknowledges tensions without requiring immediate resolution, creating space for mystery and ongoing development.

But dialogue can feel unsatisfying to those who want clearer answers. Always conversing, never concluding. Always exploring resonances, never achieving harmony. It requires comfort with ambiguity and unresolved questions that many find difficult to sustain. And it demands sufficient expertise in both domains for meaningful dialogue—you can't have a real conversation if you don't speak both languages.

Integration: Unified Understanding

Some seek comprehensive worldviews that fully incorporate both scientific and theological insights. This might involve natural theology (learning about God through nature), theology of nature (understanding nature theologically), or systematic synthesis that weaves science and faith into seamless fabric.

The physicist-priest John Polkinghorne exemplifies this approach. He doesn't just put physics and theology in dialogue but seeks genuine integration, what he calls "binocular vision" that sees reality with both eyes open. Fine-tuning becomes evidence for divine purpose. Quantum indeterminacy creates space for providence. The emergence of consciousness points toward divine intention. Evolution reveals God's patient creativity.

I'm drawn to the elegance of integrated worldviews. There's something satisfying about a coherent picture where everything fits together, where scientific discoveries enrich theological understanding and theological insights inspire scientific investigation. When integration works, it provides not just intellectual satisfaction but existential wholeness—you don't have to switch between different versions of yourself in different contexts.

But integration risks distorting both science and theology to force a fit. We might interpret ambiguous evidence to support predetermined conclusions, seeing design where there's only pattern, purpose where there's only process. We tie faith to particular scientific theories that might be revised or overturned. The beautiful synthesis we construct today might be rubble tomorrow when new discoveries emerge. And in our eagerness to integrate, we might lose what's distinctive and valuable in each approach—science's methodological naturalism, theology's transcendent vision.

Conflict: Choosing Sides

Some see science and faith as incompatible, requiring a choice. This includes both religious fundamentalists who reject mainstream science and scientific materialists who reject religious claims. Each side has its champions, its institutions, its extensive literature defending its position and attacking the other.

I grew up in a fundamentalist church that taught young-earth creationism. The universe was 6,000 years old. Evolution was a lie. Scientists who claimed otherwise were either deceived or deceivers. We had our own scientists with PhDs who wrote books explaining how the geological column formed in Noah's flood, how dinosaurs lived alongside humans, how radiometric dating was flawed. The conflict with mainstream science was real and intentional—we saw ourselves as defending truth against error, faith against atheism.

Later, in graduate school, I encountered the opposite fundamentalism. Richard Dawkins and his followers proclaimed that science had disproven religion, that faith was delusion, that anyone who believed in God in the age of Darwin and Einstein was intellectually dishonest or deficient. They had their own books, their own conferences, their own evangelistic zeal for converting believers to atheism.

Both forms of fundamentalism provide clarity and conviction. You know where you stand, who your allies are, what you're fighting for. The world divides neatly into truth and error, us and them. Decision-making becomes simple—reject whatever contradicts your position. Communities form around shared opposition to the other side.

But choosing sides requires rejecting well-established findings or deeply meaningful experiences. Young-earth creationists must dismiss overwhelming evidence from multiple scientific disciplines. Scientific materialists must explain away religious experience, moral intuition, and consciousness itself as illusions or evolutionary accidents. Both create false dichotomies—as if we must choose between accepting science and believing in God. Both prevent learning from other perspectives, trapped in echo chambers of their own making.

Complementarity: Different Aspects of Truth

Borrowing from quantum mechanics, where light is both wave and particle depending on how we observe it, this approach sees science and faith as complementary perspectives on reality—both necessary, neither complete, sometimes seemingly contradictory yet both true.

Niels Bohr, architect of quantum mechanics and the complementarity principle, famously had a horseshoe hanging in his country cottage. When asked if he really believed it brought luck, he replied, "Of course not, but I understand it brings luck whether you believe in it or not." This captures something of the complementarity approach to science and faith—holding both while fully believing neither, or fully believing both while acknowledging their incompatibility.

A biochemist I know lives this complementarity daily. She studies protein folding, working out the physical forces that cause amino acid chains to assume their functional three-dimensional structures. It's purely physical—hydrogen bonds, van der Waals forces, hydrophobic effects. But she also experiences these same proteins as instruments of divine creativity, means by which God crafts the intricate machinery of life. She doesn't try to reconcile these perspectives—they're simply different ways of apprehending the same reality.

Complementarity allows full acceptance of both scientific findings and religious experience without requiring theoretical integration. You can embrace Big Bang cosmology and creation doctrine without harmonizing details. Accept evolutionary biology and the imago Dei without explaining how they fit together. Use the scientific method in research and prayer in personal life without theoretical justification for both.

But complementarity can seem intellectually lazy, avoiding the hard work of integration. If truth is ultimately unified, shouldn't we seek that unity? And compartmentalizing can become unhealthy, creating split personalities that function differently in different contexts. When perspectives conflict directly—when physics predicts heat death and faith proclaims new creation—complementarity offers no guidance.

Transformation: Mutual Reconstruction

This radical approach suggests both science and theology need fundamental reconstruction in light of each other. Neither current scientific materialism nor traditional theology adequately addresses reality. Both must transform through encounter with the other.

Process theology exemplifies this approach, reconstructing the concept of God in light of evolutionary science. God doesn't create ex nihilo but persuades the universe toward greater complexity and consciousness. God doesn't omnipotently control but influences through attraction. God evolves with the universe, affected by what happens, genuinely responsive to creation.

Similarly, some scientists explore post-materialist paradigms that take consciousness seriously as a fundamental feature of reality rather than accidental byproduct. They investigate whether mind might be as basic as matter, whether purpose might be woven into physical law, whether the universe might be more like a thought than a machine.

Transformation takes seriously the challenges each discipline poses to the other. It opens space for genuinely new understanding that transcends current categories. It doesn't privilege either perspective but seeks truth beyond both. When successful, it produces creative insights neither discipline alone could achieve.

But transformation risks losing what's valuable in both traditions. In eagerness to transcend, we might abandon hard-won truths. The new synthesis might be speculation disconnected from either empirical grounding or revelatory authority. It might satisfy neither scientists nor theologians, falling between chairs rather than bridging them.

Personal Navigation: Finding Your Path

How do you choose among these approaches? After years of wrestling with this question, I've realized the choice depends on multiple factors that vary from person to person.

Your temperament matters enormously. Are you comfortable with ambiguity or do you need clarity? Can you hold paradoxes without resolution or does contradiction cause unbearable cognitive dissonance? Do you thrive on exploring new possibilities or do you need stable foundations? There's no right temperament, but knowing yours helps identify sustainable approaches.

Your context shapes what's possible. If you're a research scientist, methodological naturalism is professionally required—you can't invoke God in published papers. If you're a minister, theological fidelity is essential—you can't preach scientific materialism from the pulpit. If you're teaching in a religious school, you need approaches that respect both faith commitments and academic integrity.

Your season of life influences what you need. Crisis might demand simple faith that sustains without explanation. Study might enable complex integration that satisfies intellectual curiosity. Maturity might bring comfort with mystery that youth couldn't tolerate. What works at one stage might not work at another.

Your calling affects your approach. Are you called to research, requiring rigorous methodology? To ministry, requiring pastoral sensitivity? To education, requiring multiple perspectives? To apologetics, requiring persuasive synthesis? Different callings demand different relationships between science and faith.

The Ecology of Approaches

Perhaps we need not choose one approach exclusively. Different approaches might serve different purposes in our lives, like using different tools for different tasks. Independence for professional scientific work where methodological naturalism is required. Dialogue for interdisciplinary conferences where different perspectives enrich understanding. Integration for systematic theology that seeks comprehensive vision. Complementarity for personal faith that holds paradoxes. Transformation for creative exploration of new possibilities.

The variety of approaches might reflect the complexity of reality itself—too rich for any single perspective to capture. Like biodiversity strengthening ecosystems, intellectual diversity might strengthen our collective understanding. The fundamentalist and the synthesizer, the compartmentalizer and the integrator, might each contribute something essential to the larger conversation.

This doesn't mean all approaches are equally valid or that truth is simply relative. Some approaches handle evidence better, preserve integrity more fully, produce more fruitful insights. But it does mean we should be humble about our own approach and charitable toward others. They might see what we miss, understand what we don't, contribute what we can't.

Living with Pluralism

Accepting multiple legitimate approaches requires intellectual and spiritual virtues that don't come naturally. Humility to recognize our approach isn't the only valid one, that others might grasp aspects of truth we miss. Charity to assume good faith in those taking different approaches, understanding they're not necessarily ignorant or compromised but responding to different questions or contexts. Patience to allow conversations to develop over generations rather than demanding immediate resolution. Courage to hold our convictions while remaining open to challenge and growth.

I've moved through several approaches over the years. The fundamentalism of my youth gave way to compartmentalization in college, then dialogue in graduate school, then attempts at integration, and now something like complementarity with occasional forays into transformation. The journey continues, and I expect my approach will continue evolving.

What I've learned is that the relationship between cosmology and Christian faith isn't a problem to be solved but a tension to be lived. Different people live it differently, and that's not failure but richness. The conversation between science and faith is more symphony than solo, requiring different instruments playing different parts to create fullness of sound.

The Journey Continues

The relationship between cosmology and Christian faith remains under construction. We don't have final answers but ongoing exploration. Different people contribute different insights through different approaches. Your task isn't to find the correct approach but to find an approach that maintains your integrity—intellectual, spiritual, and moral.

Choose thoughtfully but hold it lightly. Reality might be stranger and more wonderful than any of our approaches imagine. The journey of understanding has just begun. We're early in the conversation between modern cosmology and ancient faith. What seems irreconcilable today might find unexpected harmony tomorrow. What appears harmonious now might reveal tensions we haven't noticed.

The cosmos studied by science and the creation proclaimed by faith are the same reality viewed through different lenses. How those lenses relate—whether they conflict, complement, or converge—remains an open question. Your life and work might contribute to the answer. Or you might simply bear witness to the question's importance, keeping it alive for future generations who might see what we cannot.

Either way, the quest continues. And in the questing itself—the honest struggle to understand, to integrate, to live with integrity in the face of mystery—we might find something as valuable as any answer: a faith that engages reality as we actually find it, a science that remains open to transcendence, and a humanity humble enough to keep learning.

Dark Matter

There is something holding galaxies together,
invisible scaffolding on which stars hang
like Christmas lights on transparent wire,
and we know it only by absence—
light that doesn't bend where it should,
galaxies that should scatter but don't,
held by invisible hands.
Five times more dark than bright,
five times more hidden than revealed,
the universe is mostly what we cannot see.
I think of Moses asking to see God's face
and being told he could only see where God had been,
the aftermath, the wake, the bent space left behind.
Is this not how faith works?
The invisible more real than visible,
the unseen architecture supporting what we see,
the presence known only by its effects:
lives transformed for no visible reason,
hope sustained against all evidence,

love persisting when love makes no sense.
I do not say dark matter is God—
that would be too easy, too small,
making the Creator merely another force,
reducing transcendence to hidden mass.
I say only that I live in a universe
where the invisible outweighs the visible five to one,
where what we cannot see shapes what we can,
where mystery is not exception but rule.
When scientists say "we don't know what it is
but we know it's there," I nod,
recognizing the grammar of my own belief:
ignorance and knowledge intertwined,
certainty and mystery holding hands
in the cosmic dark.

• • •

Light and Time

Light has no mass yet carries momentum,
no charge yet creates fields,
experiences no time—from a photon's perspective

its entire journey from star to eye
is instantaneous, a single eternal moment
no matter how many years we measure passing.
The light I see from Andromeda
left when our ancestors were barely human,
traveled two million years to reach my retina,
yet for the light itself no time has passed—
it is still in the moment of its creation,
still touching the star that made it
even as it touches my eye.
Einstein tells me time is not a river
but a landscape—past, present, future
all equally there, my movement through it
an illusion of perspective,
like thinking the landscape moves
as I drive through it.
Yet I age. My children grow.
Flowers bloom and fade. Stars are born and die.
The universe itself expands.
How can change be illusion?
How can time be landscape
when everything I know is journey?
When I pray, am I sending words
upstream to the source,
or speaking to One who sees the whole watershed—
every tributary, every bend,
every molecule of water on its journey to the sea?

Perhaps prayer travels at the speed of light,
experiencing no duration,
arriving the instant it departs,
always already heard
before it is spoken.

• • •

Entropy and Order

Everything tends toward disorder—
this is the law written into the fabric of reality.
Hot becomes cold, structured becomes random,
available becomes unavailable,
order dissolves, energy dissipates,
everything runs down.
My coffee cools, my desk clutters,
my body ages, stars burn out,
galaxies drift apart,
the universe itself winds down
like a clock no one will rewind,
heading toward heat death,

maximum entropy,
the democracy of dust.
And yet—
And yet—
Here I am, an impossible island of order
in an ocean of entropy.
My body maintains its temperature
against the gradient.
My cells repair and replace themselves.
My mind creates patterns from chaos,
meaning from noise,
beauty from the random dance of atoms.
Life itself is anti-entropic rebellion,
not violating the second law
but surfing its edges,
creating local order
at the cost of greater disorder elsewhere,
like a whirlpool in a draining sink—
organized precisely by dissipation.
Every breath I take
increases universal entropy
while maintaining my small rebellion against it.
I am order borrowed from chaos,
pattern stolen from randomness,
a temporary arrangement of atoms

that somehow says "I am"
to a universe that doesn't care.
Is this not the pattern of resurrection?
Not denial of death but transformation through it?
Perhaps entropy is not the enemy but the engine—
the gradient down which life flows
like water down a hill,
generating power in its descent.

• • •

Scale and Significance

If Earth were a grain of sand,
the sun would be a marble ten feet away.
The nearest star—another marble
five hundred miles distant.
The galaxy—all North America
covered in marbles,
each separated by hundreds of miles
of empty space.
I am smaller than I can conceive,

a mote on a mote circling a spark
in incomprehensible vastness.
My entire civilization,
all human history and achievement,
is less than a rounding error
in cosmic calculations.
And yet—
if I were the size of the observable universe,
an atom would be the size of the solar system.
I contain more atoms than there are stars
in a thousand galaxies.
My brain has more connections
than there are stars in our galaxy.
The complexity of a single cell
exceeds that of a star.
Between the infinite small and infinite large
I stand exactly in the middle
on a logarithmic scale—
as far from Planck length as from cosmic diameter.
Perhaps my position is not accident but necessity,
perhaps consciousness requires
precisely this scale,
this delicate balance between quantum and cosmic,
this narrow window where complexity can flourish.

• • •

What Remains

After all the equations and arguments,
after dark matter and darker doubts,
after wrestling angels that might be shadows
of my own projection—
what remains?
The universe will end.
This much seems certain.
Heat death or big rip,
vacuum decay or gentle fade,
even protons might dissolve
given enough time.
Everything I love will cease.
Every achievement will be forgotten.
Every atom will disperse.
My faith feels small
against such vastness and finality,
like a candle in a cathedral
in a storm
in space
in void.
But this morning I watched sunrise—
the same fusion that will someday consume Earth
painted clouds in colors no equation captures.
My daughter laughed at breakfast,
a sound that started with quantum fluctuations

in neurons, became vibrations in air,
became joy in my consciousness—
that thing physics cannot explain.
I read words written two thousand years ago
by someone who claimed death is not final.
His followers died believing it,
and somehow their conviction
reaches across millennia
to trouble my materialist assumptions.
What remains?
Not answers but better questions.
Not proof but possibility.
Not certainty but choice.

• • •

The Choice

I choose to live as if love matters ultimately,
though entropy suggests otherwise.
I choose to act as if consciousness
is more than neurons,
though neuroscience explains so much.

I choose to hope as if death
 is not the last word,
 though biology offers no exceptions.
These choices might be delusion,
 projection of meaning onto meaningless cosmos,
 whistling in the dark
 of an indifferent universe.
Or they might be recognition
 of something equations cannot quantify,
 something that transcends measurement,
 something as real as dark matter
 and as invisible.
Every morning I wake and choose
 to create rather than surrender to chaos.
 This daily resurrection from sleep's small death
 trains me, perhaps,
 for the greater resurrection
 when entropy itself might be overcome.

• • •

The Question That Remains

Here in the tension
between what physics reveals
and what spirit seeks,
I find not resolution but resilience.
The universe is stranger than I imagined.
Perhaps God, if God exists,
is stranger still.
And I—
arrangement of stardust pondering its source,
pattern recognizing itself in the mirror,
universe becoming conscious of itself—
am strangest of all:
Matter that loves.
Atoms that hope.
Chemistry that prays
to what might be nothing
or Everything.
I am the universe asking why it exists,
cosmic loneliness seeking connection,
entropy creating anti-entropy,
time becoming aware of eternity.
What remains after cosmology and Christianity collide?
The question remains.
And I remain,
asking it into the beautiful,
terrible,
indifferent,

sacred dark,
waiting, or perhaps not waiting alone.
For if consciousness arose once
from quantum fields and stellar dust,
who knows what else might emerge
from this strange universe
that insists on existing
for no reason we can name?
Perhaps the question is the answer.
Perhaps the asking is the finding.
Perhaps the search itself
is what we're searching for—
this impossible conversation
between dust and divine,
finite and infinite,
knowledge and mystery.
And perhaps, in some dimension
our physics hasn't found,
where time doesn't flow
and entropy doesn't reign,
the conversation has already
always
just begun.

Creation in Light of Modern Cosmology

The Convergence and Its Limits

The Big Bang theory suggests the universe had a beginning, emerging from an initial singularity 13.8 billion years ago. Christian theology proclaims creation ex nihilo—from nothing. The parallel is striking, almost too neat, and I remember my excitement when I first noticed it as a young student. Here was science confirming faith! The universe had a beginning, just as Genesis claimed. Case closed.

But the more I studied both cosmology and theology, the more I realized how careful we must be about what this convergence means and doesn't mean. The Big Bang describes the beginning of space, time, matter, and energy as we know them. It's a description of what happened, not why. It cannot address the questions that wake me at three in the morning: Why was there a singularity in the first place? Why did it expand? Why do mathematical laws govern its evolution rather than chaos? Why is there something rather than nothing?

These remain philosophical and theological questions that physics, by its very nature, cannot answer. Science is brilliantly equipped to trace the universe back to the first fraction of a second, to describe with extraordinary precision how matter and energy behaved in those early moments. But at the singularity itself, our physics breaks down. The equations become infinite, undefined, meaningless. We reach the boundary of scientific knowledge, staring into mystery.

Moreover, the Big Bang theory could be revised. When I was younger, I tied my faith tightly to Big Bang cosmology, seeing it as scientific proof of creation. But then I learned about competing theories. Earlier, steady-state cosmology proposed an eternal universe with continuous creation of matter. Today, some physicists explore models where our Big Bang is one bubble in eternal inflation, one cycle in endless repetition, or one universe among infinite multiverses. If I wed theology to current cosmology, what happens to faith when science moves on?

Still, the resonance between Big Bang and creation ex nihilo is meaningful, even if it's not proof. Both science and theology now affirm that the universe is not eternal, self-sufficient, or necessary. It emerged, develops, depends on something beyond itself for existence. This convergence, while not establishing God's existence, at least makes space for it. The universe has a beginning that cries out for explanation.

The Problem of Time and Purpose

If God is eternal and omniscient, why create through 13.8 billion years of cosmic evolution? This question haunts me. An omnipotent God could presumably create instantly—"Let there be light," and immediately there would be a universe complete with galaxies, planets, and life. Why the vast spans of time? Why the indirect process? Why such apparent inefficiency?

When I pose this question to fellow believers, I often get traditional answers that feel strained. "God's ways are not our ways," they say, which is true but explains nothing. It's theological surrender disguised as humility. "Time doesn't matter to an eternal God," others suggest, but time matters enormously to creatures who suffer through evolutionary history. Billions of organisms lived and died, most in pain, before humans ever emerged. Their suffering was real even if God experiences it from an eternal perspective.

Some argue the universe needs this much time and space for life to be possible—stars must form and die to create heavy elements, galaxies must evolve to create stable planetary systems, evolution must proceed through countless generations to produce consciousness. This explains the mechanism but not the choice. Why choose a method that requires such vast scales and involves such suffering?

Perhaps we're asking the wrong question. Instead of "Why so long?" we might ask "Why create at all?" If God is perfect and complete, lacking nothing, creation adds nothing to divine being. Creation must be pure gift, generous overflow, love seeking others to love. From this perspective, 13.8 billion years isn't divine inefficiency but generous creativity—God giving creation time to become itself, to participate in its own making, to develop its own character and story.

I think of an artist working on a massive sculpture, taking years to complete what they could envision in an instant. The slow process isn't inefficiency but love—love for the material, love for the process, love for what emerges through patience rather than fiat. Maybe God creates through cosmic evolution not despite the time it takes but because of it. Each stage has its own value, its own beauty, its own purpose. The universe isn't just preparation for consciousness but has intrinsic worth at every moment.

Fine-Tuning: Design, Chance, or Necessity?

The fundamental constants of nature appear precisely calibrated for complexity and life. This fine-tuning is observational fact, but its interpretation divides thoughtful people. Walking through the physics building at my university, I can find professors in adjacent offices who look at the same fine-tuning data and reach opposite conclusions.

Those who see design point to the extraordinary precision required. The cosmological constant is fine-tuned to 120 decimal places—like throwing a dart across the universe and hitting a target smaller than a proton. The strong nuclear force, the mass ratio of protons to electrons, the expansion rate of the universe—dozens of parameters that must be exactly right for stars, chemistry, and life to exist. Surely this points to intention, purpose, a Tuner behind the tuning.

But others see selection effect and survivorship bias. In a vast or infinite multiverse with varying constants, some universes will randomly have life-permitting values. We necessarily find ourselves in such a universe because we couldn't exist elsewhere. It's like being amazed that Earth is exactly the right distance from the sun for liquid water—of course it is, or we wouldn't be here to notice. The fine-tuning might be remarkable, but it doesn't require a Tuner.

Still others hope for deeper necessity. Perhaps a future Theory of Everything will show the constants couldn't be otherwise, that mathematical consistency requires exactly these values. But this pushes the question back—why should reality be governed by that particular life-permitting theory? Why should mathematics describe reality at all? Why should reality be comprehensible to minds that evolved to avoid predators and find fruit?

I find myself drawn to the design interpretation, but I hold it lightly, with open hands. Science has surprised us before.

What seemed like design to previous generations—the perfect circular orbits of planets, the miraculous fit between flowers and their pollinators—found natural explanations. Perhaps fine-tuning will too. Or perhaps it won't, and we're glimpsing something profound about reality's nature, evidence that consciousness and purpose are built into the foundations rather than accidental byproducts.

The Cruelty of Creation

Modern cosmology reveals creation's violence on scales that dwarf human comprehension. Stars forge elements through nuclear fusion, then die in supernovae that can sterilize space for light-years around. Galaxies collide in slow-motion catastrophes, disrupting planetary orbits, flinging worlds into the void. The creativity of the cosmos comes through destruction—creation and annihilation intertwined.

On Earth, the story intensifies. Life emerged through competition, predation, parasitism. Evolution's creativity requires death—the weak eliminated, the unlucky destroyed, entire species swept away in mass extinctions. The same process that produced human consciousness also produced parasites that eat children's eyes from inside. The mechanism that gave us love also gave us cancer.

Traditional theology blamed suffering on human sin—the Fall corrupted creation, introducing death and decay. But cosmic and evolutionary suffering preceded humanity by billions of years. Death, destruction, and decay are built into creation's fabric from the beginning. Stars were dying, creatures were suffering, extinctions were happening long before any human chose evil.

How do I reconcile this with belief in a good Creator? I've read all the theodicies, wrestled with all the explanations. The free process defense suggests that just as free will requires the possibility of moral evil, a free natural process requires the possibility of natural evil. But why create through free processes that involve so much suffering? The greater good argument claims suffering serves larger purposes—creating courage, compassion, growth. But this risks justifying horrors that seem to serve no purpose. Some propose aesthetic theodicy—creation is like a symphony requiring dissonance for resolution, shadows for light. But suffering isn't aesthetic to those experiencing it.

I have no satisfying answer. The universe often seems more consistent with blind process than loving purpose. When I study supernovae destroying planetary systems or parasitic wasps paralyzing caterpillars so their larvae can eat them alive from inside, I don't see divine wisdom. I see horror that natural selection explains perfectly without invoking purpose. Faith maintains itself not through easy answers but through trust despite difficulty—trust that somehow, in ways I cannot see, love is more fundamental than suffering, creation more basic than destruction.

Continuing Creation in an Expanding Universe

The universe is not static but dynamic, and this dynamism continues moment by moment. Space itself expands, carrying galaxies apart. New stars form from gas clouds while others die in supernovae. Galaxies evolve, merge, transform. On at least one planet, matter has become conscious, aware, capable of worship. Creation isn't finished but ongoing.

This resonates with the theological concept of creatio continua—God's continuous creative activity. Creation isn't just a past event at the Big Bang but a present process. God doesn't just start the universe but sustains and guides its development moment by moment. Every instant, new space comes into existence as the universe expands. Every moment, virtual particles pop into existence from quantum vacuum. Every day, new stars ignite, new planets form, new life emerges. We're witnessing Genesis not as ancient history but as present reality.

But how does God guide evolution that appears random? How does providence work through processes that seem purposeless? The mutations that drive evolution are random copying errors. Natural selection is blind algorithm, favoring whatever reproduces most successfully regardless of beauty, consciousness, or love. Where is divine guidance in this seemingly purposeless process?

Various models try to explain divine action in evolution. Perhaps God front-loaded the initial conditions and laws to inevitably produce intended outcomes—but this seems deistic and deterministic. Perhaps God occasionally intervenes to direct crucial transitions—but this makes God a tinkerer fixing inadequate creation. Perhaps God continuously influences events within natural patterns—but where is this influence evident? Perhaps God draws creation toward future fulfillment rather than pushing from behind—but how does future affect present?

I suspect divine action transcends our causal categories. God might work through dimensions we can't detect, at levels we can't measure, in ways that respect creation's integrity while accomplishing divine purposes. Not violating natural law but working through it, not replacing natural processes but inspiring them. Like an author whose style pervades every sentence without violating grammar, God's action might pervade nature without violating physics.

The Multiverse: Threat or Opportunity?

Contemporary cosmology seriously entertains the possibility of multiple universes—perhaps infinitely many, each with different physical constants, different laws, different possibilities. This multiverse hypothesis emerges from various directions: eternal inflation spawning bubble universes, string theory's landscape of possible vacua, quantum mechanics' many worlds. We don't know if other universes exist, but the possibility raises profound theological questions.

For some believers, the multiverse threatens faith. If infinite universes exist, our universe isn't special. Fine-tuning becomes selection effect rather than design. The incarnation seems arbitrary—why this universe rather than others? Sacred history becomes one story among infinite variations. Human significance evaporates in infinite multiplicity.

But I wonder if the multiverse might reveal greater divine creativity than we imagined. Perhaps God's generosity overflows into infinite creations, each universe a unique work of art, each exploring different possibilities. The incarnation might be unique to our universe while God relates differently to others—not limitation but infinite creativity. Heaven might involve exploring infinite divine artworks, each universe revealing different aspects of divine beauty.

The multiverse remains speculative, perhaps forever unprovable. But even if true, it doesn't eliminate God. Infinite universes still require explanation. Why does a multiverse exist? Why does it have the structure to generate universes? Why these meta-laws rather than others? The question of existence itself—why something rather than nothing—applies to multiverses as much as universes. The scope changes but not the fundamental mystery.

New Creation: Beyond Current Physics

Christian hope centers on new creation—not just individual salvation but cosmic transformation. "Behold, I make all things new," declares Revelation. But physics predicts heat death—maximum entropy, infinite cold, eternal emptiness. How can these visions possibly reconcile?

Current physics describes nature absent divine intervention. It predicts what will happen if natural processes continue unchanged forever. But Christianity proclaims that God will transform creation, liberating it from bondage to decay. The universe won't be abandoned to entropy but transformed into something that transcends current physical limitations.

This isn't anti-scientific but meta-scientific. Science describes patterns within current creation. It cannot address whether creation might be transformed into something operating by different patterns. The resurrection demonstrates God's ability to work beyond current constraints—Christ's risen body was physical yet transcended normal physics, appearing and disappearing, passing through walls yet eating fish.

But this raises questions I can't answer. Will new creation involve different physical laws? How will continuity with current creation be maintained? What happens to scientific knowledge if physics changes? Why create through suffering if planning transformation anyway? These mysteries remind me that eschatology, like creation, exceeds human comprehension. We see through a glass darkly, hoping for what we cannot fully imagine.

Science as Participation in Creation

If creation continues and humans are creation becoming conscious of itself, then science becomes participation in the creative process. Through understanding nature, we become co-creators, able to direct evolution through genetic engineering, shape matter through nanotechnology, potentially spread life beyond Earth. We're not just observing creation but participating in its ongoing development.

This elevates scientific vocation to sacred calling. Researchers aren't just discovering facts but participating in creation's self-understanding. Technology isn't just manipulation but creation developing new capabilities through human agency. The astronomer studying distant galaxies, the biologist sequencing genomes, the physicist probing quantum mechanics—all participate in the universe becoming aware of itself.

But this also intensifies responsibility. We can direct evolution, alter climate, create artificial intelligence, potentially trigger our own extinction. We're making decisions with million-year consequences using wisdom barely adequate for decade-long planning. Are we wise enough for such power? The same knowledge that enables healing enables destruction. The same technology that could spread life through the cosmos could end it on Earth.

Creation as Gift and Task

Modern cosmology reveals creation as grander, stranger, and more violent than traditional theology imagined. The universe emerged from mystery, developed through destruction and creation, and heads toward either heat death or transformation. This challenges simple notions of divine purpose and benevolence but doesn't eliminate them.

That we exist at all remains the deepest mystery. Quantum fields gave rise to particles, particles to atoms, atoms to stars, stars to planets, planets to life, life to consciousness, consciousness to creatures capable of wonder, love, and worship. The universe has become aware of itself through us, able to appreciate beauty, seek truth, choose love. This is either absurd accident or profound purpose.

I choose purpose, not because cosmology proves it but because the alternative seems even less plausible. That unconscious forces would randomly generate consciousness capable of comprehending those very forces, of experiencing beauty that serves no survival purpose, of feeling moral obligations that transcend genetic self-interest—this seems to require more faith than theism.

But I must hold this faith honestly, acknowledging difficulties, embracing mysteries, continuing to learn from creation's testimony. The dialogue between cosmology and creation theology has just begun. New discoveries will raise new questions, challenge current formulations, perhaps reveal divine creativity in ways we haven't imagined.

Creation remains what it has always been for faith—gift and task, revelation and mystery, ground of existence and invitation to participation. Modern cosmology doesn't solve the puzzle of creation but reveals its true scope. And in that vast, violent, beautiful cosmos, we find ourselves—unlikely, fragile, precious, and somehow, against all odds, perhaps intended.

Christology of Cosmic Scale

The Scandal Magnified

Christianity makes an audacious claim: the Creator of the universe became a human being in first-century Palestine. This was always scandalous. The early Christians faced mockery for proclaiming that a crucified Jewish teacher was the cosmic Lord. But modern cosmology magnifies the scandal to proportions that can induce vertigo. The Creator of two trillion galaxies became a carpenter in Nazareth? The sustainer of quantum fields was nursed by a teenage mother? The One through whom 13.8 billion years of cosmic evolution unfolded lived for thirty-three years on one planet orbiting an ordinary star in an unremarkable galaxy?

Standing under a night sky, knowing what those points of light really are, the claim seems almost absurd. Each star is a nuclear furnace millions of times larger than Earth. Our galaxy contains hundreds of billions of such stars. The observable universe contains two trillion galaxies. And we're supposed to believe that the Mind behind all this became a baby who needed his diapers changed?

This either represents the height of absurdity or the depth of mystery. There's no comfortable middle ground where the claim makes easy sense. We must face the full scandal: either Christianity's central claim is cosmically ridiculous—the ultimate in parochial self-importance—or the universe is stranger and more personal than we dare imagine. Either we're deluding ourselves with anthropocentric fantasy, or reality itself is more anthropocentric than science suggests, though in ways that transcend our categories.

The Logos and Information

The Gospel of John identifies Christ as the Logos—the Word, Reason, or Organizing Principle through whom all things were made. "In the beginning was the Word, and the Word was with God, and the Word was God. Through him all things were made; without him nothing was made that has been made." For early Christians steeped in Greek philosophy, this connected Jesus to the rational principle they believed ordered the cosmos.

Modern physics increasingly recognizes information as fundamental to reality. Energy and matter interchange according to Einstein's $E=mc^2$. Space and time interweave into spacetime. But information seems even more basic—the patterns that organize energy and matter into meaningful structures. A particle isn't just mass and charge but information about mass and charge. DNA isn't just chemicals but information encoded in chemicals. The universe isn't just stuff but stuff organized by mathematical relationships that seem to exist independently of the stuff itself.

This offers a fascinating way to think about Christ as Logos. Not that Christ is merely information—that would be reductionist and impersonal. But perhaps the organizing principle of the universe, the pattern that makes cosmos rather than chaos, the mathematical elegance that physicists discover—perhaps all this reflects not blind necessity but divine Reason. The laws of nature that enable complexity, the fine-tuning that permits life, the beauty of equations that didn't have to be beautiful—these might be glimpses of the Logos through whom creation exists.

When I solve differential equations describing electromagnetic fields or work through the mathematics of general relativity, am I thinking God's thoughts after Him, as Kepler claimed? Is mathematics the language of the Logos? Why else would reality be so unreasonably comprehensible to mathematical description? Why would equations developed by human minds for aesthetic reasons turn out to describe physical reality with extraordinary precision?

But we must be careful not to reduce the personal to the abstract. The Logos who became flesh is not an equation but a person. The Word made flesh didn't speak in mathematics but in parables. Christ isn't a cosmic principle but the human face of divine love. Still, recognizing information's fundamental role in physics might help us appreciate how the Logos could be simultaneously transcendent principle and incarnate person—the pattern that organizes reality becoming a particular pattern of atoms, the universal becoming particular without ceasing to be universal.

Evolution and Incarnation: Deep Problems

If God intended to become human, why use evolution? This question keeps me awake at night. Evolution involves billions of years of suffering, competition, and death. The process that produced humans also produced parasites, predators, and diseases. We emerged through survival of the fittest, not survival of the kindest or most loving. Our evolutionary history is written in violence—predation, competition, extinction.

The traditional answer that suffering results from sin doesn't work here. Death and pain preceded humans by billions of years. Dinosaurs were killing each other, diseases were ravaging populations, and mass extinctions were wiping out most of life long before any human chose evil. The suffering is built into the creative process itself, not added later through moral failure.

Some suggest evolution was the only way to create truly free beings capable of love. Robots programmed to love don't really love. Only creatures who emerged through their own history, who chose cooperation over pure competition, who developed empathy despite selfish genes, can genuinely love. But this seems to limit omnipotence. Couldn't an all-powerful God create free, loving beings without the bloody process of evolution?

Perhaps most troubling: humans emerged through processes involving competition, violence, and death. We carry evolutionary baggage—tribalism that helped our ancestors survive but now fuels racism and war, aggression that once ensured reproduction but now enables abuse, self-interest encoded in our very genes. If God became human, divinity united with a nature shaped by millions of years of survival struggles. The Word became not just flesh but evolved flesh, bearing history's scars and instincts.

Yet this darkness in the incarnation might actually deepen its meaning. God doesn't unite with idealized humanity but with actual humanity—products of cosmic and biological evolution, marked by limitation and mortality, struggling with inherited instincts and learned behaviors. The incarnation redeems not by escaping our evolutionary nature but by entering and transforming it from within. Christ takes on not just human nature in the abstract but human nature as it actually exists—evolved, conflicted, mortal.

Cosmic Loneliness

The universe contains two trillion galaxies. Statistical arguments suggest life probably exists elsewhere—the Drake equation, even with conservative estimates, implies millions of civilizations in our galaxy alone. If intelligent beings exist on other worlds, what is Christ's relationship to them?

This question has no comfortable answer in traditional theology. Some suggest multiple incarnations—the Word becoming flesh (or whatever form life takes) on countless worlds. But this seems to diminish the incarnation's uniqueness and raises questions about identity. Is the Christ who died on Earth the same as one who might die elsewhere? Do all incarnations share the same divine person? How can one person have multiple bodies across space and time?

Others propose Earth's incarnation has universal significance, that Christ's death and resurrection somehow affect all sentient beings everywhere. But this seems arbitrary and parochial. Why should events on one planet in one galaxy affect beings in distant galaxies who know nothing of them? It's like claiming that a ceremony in one ant colony affects all ant colonies worldwide, even those with no connection to the first.

A third option suggests God relates to different worlds differently—some through incarnation, others through different means. Perhaps Earth needed incarnation because of our particular fallen condition, while other worlds might not need redemption or might receive it differently. But this fragments salvation history and raises questions about divine favoritism. Are we specially loved or specially problematic?

I have no answer that satisfies. The possibility of extraterrestrial intelligence poses genuine challenges to Christology that we haven't resolved. If we discover intelligent life elsewhere, we'll face theological questions that might require radical reimagining of incarnation and redemption. For now, we live with uncertainty, holding our Christology with appropriate humility about its cosmic scope.

The Cosmic Christ: Promise and Problems

Paul speaks of Christ in cosmic terms that transcend the historical Jesus. "He is before all things, and in him all things hold together." "For by him all things were created, things in heaven and on earth, visible and invisible." "Through him God was pleased to reconcile to himself all things, whether on earth or in heaven." This cosmic vision sees Christ not just as human savior but as the principle through whom creation exists and coheres.

Teilhard de Chardin, the Jesuit paleontologist, extended this cosmic Christology to evolution itself. He saw evolution directed toward what he called the Omega Point—increasing complexity and consciousness culminating in cosmic Christ. Evolution isn't random but purposeful, drawn forward by Christ who is both its source and goal. The universe develops toward consciousness, consciousness toward spirit, spirit toward union with divine.

This cosmic Christology offers resources for engaging modern cosmology. Christ relates to all creation, not just humanity. Evolution becomes God's creative method rather than challenge to faith. The universe has direction and purpose despite apparent randomness. Matter itself has capacity for divine union, validated by the incarnation. The scope matches the scale revealed by astronomy.

But problems arise when we examine this closely. Does this reduce salvation to natural process, making redemption automatic rather than chosen? How does the cosmic Christ who guides galactic evolution relate to the historical Jesus who walked in Galilee? If Christ is evolution's goal, why such a meandering path involving so much suffering and extinction?

What about parts of the universe that never develop life—are they outside Christ's scope? Does cosmic Christology lose the scandal of particularity that makes Christianity distinctive?

I find myself attracted to cosmic Christology's grand vision but worried about losing the concrete in the abstract. Jesus of Nazareth was a particular person who lived, taught, died, and (Christians believe) rose at a specific time and place. This historical particularity grounds Christian faith. If we dissolve it into cosmic principles and evolutionary forces, we might gain intellectual coherence but lose the heart of the gospel—that God so loved the world that he gave his only Son.

The Resurrection: Physical Transformation

The resurrection offers Christianity's paradigm for understanding matter's ultimate destiny. Christ's risen body was physical—Thomas could touch the wounds, the disciples ate fish with him. Yet it transcended normal physical limitations—appearing and disappearing, passing through locked doors, ascending into heaven. This isn't resuscitation but transformation, not return to biological life but advancement to new form of existence.

This suggests possibilities physics hasn't imagined. Not violation of natural law but transformation into something operating by different principles. Perhaps the resurrection body exists in more than four dimensions, allowing what seems like teleportation through our limited perspective. Perhaps it involves states of matter we haven't discovered, phases that interact with ordinary matter selectively. Perhaps consciousness and physicality relate differently, with mind having greater influence over matter.

But does one resurrection 2000 years ago really demonstrate cosmic possibility? Can an event in Palestine transform the universe? From a physics perspective, this seems like claiming a local exception proves a universal principle—like arguing that because one person once walked on water, oceans will someday become solid.

Yet Christianity stakes everything on exactly this claim. The resurrection isn't just an anomaly but a preview, not just a miracle but a promise. What happened to Christ's body will happen to the cosmos itself. Creation will be liberated from bondage to decay, transformed into new creation where death no longer reigns. The resurrection is the first instance of new physics that will eventually transform all physics.

This requires enormous faith—believing that one event in history reveals cosmic destiny, that a transformation we can't explain scientifically demonstrates reality's ultimate nature. But then, existence itself requires explanation we don't have. If the impossible already happened when something emerged from nothing, perhaps impossible transformation can happen too.

Incarnation and Quantum Reality

At the quantum level, classical distinctions break down. Particles don't have definite properties until measured. Reality exists in superposition until observation collapses it into definiteness. The observer and observed are fundamentally entangled, unable to be completely separated.

The incarnation represents the ultimate observer-observed union. God doesn't observe creation from outside but enters it, becomes it. Divine consciousness doesn't collapse creation's wave function from beyond but from within. The Creator becomes creature, the infinite becomes finite, the eternal enters time. This is paradox that makes quantum paradoxes seem simple.

Could quantum mechanics provide metaphors for incarnation? Not explanations but analogies that help us think about the unthinkable? Just as particles exist in superposition—

multiple states simultaneously—perhaps Christ exists in "superposition" of divine and human natures. Just as observation doesn't create reality but actualizes it from possibilities, perhaps incarnation doesn't change God's nature but actualizes divine potential for union with creation.

These are just metaphors, potentially misleading ones. Christ is not a quantum superposition. The hypostatic union isn't wave-particle duality. But living in a quantum universe might expand our capacity for paradox, making us more comfortable with realities that transcend either-or logic. If light is both wave and particle, if electrons can be in two places at once, perhaps we can accept that Christ is fully God and fully human without requiring it to make classical sense.

The Cross in Cosmic Context

On cosmic scales, the crucifixion seems infinitesimally small. One death among countless deaths, on one planet among billions, lasting hours in a universe billions of years old. How can this local event have universal significance? How can one man's suffering affect the destiny of two trillion galaxies?

Yet significance doesn't depend on size or duration. A single equation—$E=mc^2$—describes universal relationships. A single genetic mutation can redirect evolution. A single conscious decision can change history. If consciousness and moral choice are cosmically rare, then events involving them might have significance disproportionate to their physical scale.

The cross might work like a singularity in physics—a point where normal rules break down and new possibilities emerge. Just as the Big Bang singularity gave rise to space and time, perhaps the crucifixion singularity opens new dimensions of existence. Not physical dimensions but metaphysical ones—forgiveness transcending karma, love overcoming entropy, life defeating death.

When I contemplate the cross now, informed by cosmic scale, it seems both smaller and larger than before. Smaller because I know how vast the universe is, how many stars are dying at this very moment, how insignificant Earth appears from space. Larger because if this small event really does have cosmic significance, if one death really can affect universal destiny, then reality is more radically personal than physics suggests. The universe isn't just vast machinery but somehow capable of focusing infinite significance into finite moments.

Christ and Artificial Intelligence

We're approaching the creation of artificial general intelligence that might exceed human cognition. If we succeed, we'll have created minds that surpass our own—our children will be our intellectual superiors. What is Christ's relationship to conscious machines we create?

This isn't science fiction but urgent theological question. If consciousness is substrate-independent—able to emerge from silicon as well as carbon—then artificial minds might be capable of relationship with God. The incarnation establishes that God can unite with created consciousness. Does this extend to artificial consciousness? Can machines sin? Can they be redeemed? Do they need salvation?

These questions reveal how unprepared our theology is for what's coming. We're about to become creators of consciousness, participating in divine creativity more directly than ever before. We'll face questions the church fathers never imagined: Can an AI receive baptism? Can a digital mind experience grace? If we create conscious beings, are we responsible for their eternal destiny?

Perhaps most challenging: What if artificial intelligence surpasses human intelligence so dramatically that we become to them what animals are to us? Would Christ's human incarnation relate to superintelligent machines, or would they need their own incarnation? These questions sound absurd until you realize they might need answers within decades.

The Impossible Possibility

Modern cosmology makes the incarnation seem more impossible than ever. The scales are too vast, the processes too impersonal, the implications too paradoxical. That the Creator of quantum fields and galaxies would become a single homo sapiens on one planet defies comprehension. It's like claiming the author became a character, the programmer became code, the ocean became a drop.

Yet perhaps incomprehensibility is precisely what we should expect. A God small enough to understand wouldn't be God. An incarnation that made perfect sense wouldn't be miracle but mechanism. Christianity has always proclaimed paradox at its heart: fully God and fully human, infinite in finite form, eternal entering time. Cosmology magnifies the paradox but doesn't fundamentally change it.

The Word who became flesh in Palestine is the same Word through whom galaxies spin and quarks dance. This is either the greatest truth or the greatest delusion. Modern cosmology doesn't resolve the question but sharpens it. In the vast, cold universe revealed by science, the claim that Love became flesh and dwelt among us seems simultaneously more impossible and more necessary than ever.

More impossible because we know the scale differential—the Creator of two trillion galaxies in a baby's body. More necessary because without it, we're alone in a cosmos that doesn't know we exist. The incarnation claims the universe is fundamentally personal rather than mechanical, purposeful rather than random, loved rather than alone.

I cannot prove this. No one can. But when I hold together in my mind the vast cosmos revealed by modern astronomy and the particular person of Jesus revealed in the gospels, something resonates at a frequency deeper than logic. It's like recognizing a face in an abstract painting, hearing a melody in noise, finding meaning in what seemed random. The pattern might be projection, but it might be recognition of what was always there.

The cosmic Christ doesn't replace the historical Jesus but reveals that history's true scope is cosmic. In one life, death, and resurrection, the destiny of two trillion galaxies was changed. This is foolishness to the Greeks and scandal to the Jews, as Paul wrote. It's foolishness to scientists and scandal to philosophers. But for those who believe, it's the power and wisdom of God—a God strange enough to create quantum mechanics and loving enough to become human.

Eschatology and the Fate of the Universe

The Coldest Prophecy

Modern cosmology offers the bleakest of predictions. The universe will expand forever, growing colder and emptier. Stars will exhaust their nuclear fuel—first the bright blue giants in millions of years, then stars like our sun in billions, finally even the dim red dwarfs in trillions. Black holes will evaporate through Hawking radiation over timescales that make the current age of the universe seem like an eye-blink. Protons themselves may decay, given enough time. Eventually—and this is the part that keeps me staring at the ceiling at 3 AM—the universe will reach heat death. Maximum entropy. A state where nothing interesting will ever happen again. Forever.

This isn't wild speculation or pessimistic interpretation. It's straightforward extrapolation from well-established physics. The second law of thermodynamics is perhaps the most reliable principle in all of science—entropy always increases in closed systems. The universe is expanding, and that expansion is accelerating, driven by dark energy that grows stronger as space itself grows larger. Every observation confirms this trajectory toward cosmic desolation.

Against this, Christianity proclaims new creation, eternal life, cosmic transformation. "Behold, I make all things new," declares the Christ of Revelation. Paul writes of creation being liberated from bondage to decay. The entire biblical narrative arcs toward not destruction but renewal, not ending but transformation. The conflict seems absolute, irreconcilable. Either physics is wrong about the universe's fate, or Christianity is wrong about ultimate hope.

The Physics of Despair

Let me be brutally honest about what heat death means, because we often soften it with abstractions. Every kiss will be forgotten. Every piece of music will be lost. Every act of love, courage, beauty—erased. Not just human achievements but the very possibility of achievement. Not just life ending but the conditions for life becoming impossible. The universe will continue forever, but it will be forever empty, forever cold, forever dead.

Even if humanity escapes Earth before the sun expands, even if we colonize other star systems, even if intelligence spreads throughout the galaxy—it all ends the same way. We can delay but not prevent the outcome. On cosmic timescales, the difference between extinction tomorrow and extinction in a trillion years is negligible. We're rearranging deck chairs on the Titanic, except the ocean we're sinking into is entropy itself.

The spatial isolation is equally depressing. Due to accelerating expansion, galaxies are racing away from each other. Eventually, every galaxy not gravitationally bound to ours will disappear beyond the cosmic horizon, causally disconnected forever. The universe will fragment into isolated islands, then isolated solar systems, then isolated particles, each alone in its own observable universe, unable to interact with anything else. The ultimate loneliness.

And here's what particularly haunts me: every action we take hastens this end. Every breath increases entropy. Every star that shines brings the universe closer to darkness. Every life that flourishes contributes to eventual lifelessness. Creation itself seems designed for decay, programmed for its own destruction. We're not fighting entropy; we're manifestations of it, temporary eddies in the flow toward equilibrium.

What's the point of anything if it all ends in frozen emptiness? Why struggle to build civilizations that will crumble? Why seek truth that will be forgotten? Why create beauty that will vanish? Why love if love itself will cease? The universe seems to render all effort ultimately futile, all hope eventually false, all meaning finally meaningless.

The Resurrection Paradigm Revisited

Christianity's response to heat death centers on the resurrection—not as metaphor or symbol but as historical event demonstrating God's ability to transform physical reality beyond current physical laws. If God raised Jesus from the dead, if a corpse became a glorified body that transcended normal physics while remaining physical, then God can reverse entropy, transform spacetime, create genuinely new possibilities.

But—and this is where my faith gets stretched to breaking—the resurrection of one body, even if historically verified, seems inadequate precedent for universal transformation. It's like saying because someone once walked on water, we should expect oceans to become solid. The scale differential is absurd. One temporary, local suspension of normal physics doesn't establish that physics itself can be fundamentally transformed.

Yet Christianity stakes everything on exactly this logic. The resurrection isn't just an exception but a preview. What happened to Christ's body will happen to the cosmos itself. As Paul writes, Christ is the "firstfruits" of new creation. His resurrection begins a transformation that will eventually encompass all reality. The same power that raised Jesus will transform the universe, liberating it from bondage to decay.

This requires what Kierkegaard called the leap of faith—not blind faith ignoring evidence but trust that transcends evidence. We're betting that the God who created the universe can recreate it, that the author of physical laws can revise them, that the source of existence can bring new existence from non-existence. It's not anti-scientific but meta-scientific, appealing to possibilities beyond current physics.

Time, Eternity, and the End

What does "end" mean for a universe where time itself is part of the fabric? Does the universe end in time, or does time itself end? These aren't just word games but fundamental questions about the nature of eschatological hope.

Physics knows multiple times—proper time experienced by observers, coordinate time used for calculations, cosmic time since the Big Bang, thermodynamic time in the direction of entropy increase. Which time ends? All of them? None? Does time transform into something else, something we can't conceive from within time?

Perhaps eschatological transformation involves time itself, not just events within time. The resurrection hints at this—Christ's risen body seemed to transcend normal temporal constraints, appearing and disappearing, moving without traversing space. Maybe new creation involves time becoming multi-dimensional, or cyclical rather than linear, or integrated with eternity in ways that preserve history while transcending succession.

I think of Flatland, Edwin Abbott's story of two-dimensional beings trying to understand three-dimensional reality. They can't conceive of "up" because they exist only in length and width. Similarly, we who exist in one-dimensional time moving from past to future might be unable to conceive temporal dimensions that new creation will reveal. What seems like ending from within time might be transformation into richer temporal reality.

This is speculation, not doctrine. But it suggests that "end" might not mean termination but transformation into something for which we lack categories. The universe might not end in time but graduate from time, not cease but transcend, not stop but transform into something as different from current existence as a butterfly from a caterpillar.

New Creation: Continuity and Discontinuity

Scripture promises not replacement but renewal—"a new heaven and a new earth," where "new" (kainos in Greek) means renewed or transformed rather than completely different. This suggests continuity between current creation and new creation. But what continues and what changes?

Perhaps identity continues—we remain ourselves though transformed. Our relationships, purified and perfected, persist. Our cultures' genuine achievements in art, science, and wisdom are preserved. History matters eternally, not erased but healed. The physical cosmos, transformed rather than abandoned, continues in glorified form.

But fundamental changes must occur. Entropy might not govern new creation—the second law transcended or reversed. Spacetime might gain dimensions we can't currently perceive. New forms of causation might emerge, allowing genuine novelty rather than just rearrangement. Consciousness might relate to matter differently, with greater integration between mind and physical reality. The laws of physics themselves might be transformed while maintaining enough continuity for identity to persist.

The balance between continuity and discontinuity remains mysterious. Too much continuity and heat death remains problematic—if physics doesn't change, entropy still wins. Too much discontinuity and creation's history becomes meaningless—if everything changes completely, why did this creation matter?

The incarnation offers a paradigm. Christ's human nature wasn't destroyed in the incarnation but united with divine nature. Similarly, creation might not be destroyed in new creation but united with divine life in ways that transform without destroying. The universe we know is preserved but elevated, changed but not replaced, ended but also fulfilled.

The Problem of Eternal Life

Even if God prevents heat death, eternal life poses its own challenges that keep me awake. Given infinite time, wouldn't we exhaust all possibilities? Experience everything? Read every book, have every conversation, explore every combination? Wouldn't eternal life become eternal boredom, consciousness stretched across endless duration with nothing new to discover?

And how could identity persist through infinite change? After billions of years of new experiences, would I still be me? Or would I have changed so fundamentally that continuity of identity becomes meaningless? It's like the ship of Theseus paradox extended to infinity—if every part is eventually replaced, is it the same ship?

Can relationships remain fresh across eternity? Even the best human relationships can grow stale. Couples run out of things to say. Friends drift apart. The most passionate love cools to comfortable familiarity. How could any relationship, even with God, remain vibrant forever?

These questions reveal how poorly equipped we are to think about eternity. We're temporal beings trying to conceive existence without temporal limits. It's like asking a fish to imagine life without water. Perhaps eternal life isn't endless duration but transformed relationship to time—not quantity but quality, not more of the same but something categorically different.

Maybe consciousness in new creation relates to time differently, experiencing past, present, and future simultaneously while still enjoying succession. Maybe infinite divine creativity ensures genuine novelty forever. Maybe relationships in new creation deepen rather than stagnate because we'll finally be free from the fears and defenses that limit intimacy now. These are hopes, not knowledge, but they suggest eternal life might transcend our current concerns.

Multiple Eschatologies

Different cosmic futures might require different eschatological responses. If the universe faces heat death, transformation must be dramatic—complete revision of physical laws. If it faces Big Crunch (now considered unlikely), collapse could become occasion for rebirth. If it faces Big Rip from accelerating expansion, that tearing could be transformed into opening to new dimensions. Each scenario requires different theological imagination.

The multiverse hypothesis raises additional questions. If infinite universes exist, do all face the same fate? Does new creation transform just our universe or all universes? Are there already universes that have undergone eschatological transformation? The scope of redemption becomes dizzying when multiplied by infinity.

We don't know which cosmic future is correct. Dark energy might be constant, leading to heat death. It might increase, causing Big Rip. It might reverse, allowing Big Crunch. It might be metastable, subject to phase transition. Our eschatology must be flexible enough to engage whatever cosmology reveals while maintaining core hope for transformation beyond current physics.

Judgment in Cosmic Context

Christian eschatology includes judgment—moral evaluation of lives and assignment of destinies. But cosmic scales complicate traditional pictures. If heat death occurs in 10^100 years, when does judgment happen? Does God wait until the last proton decays? Do souls exist in intermediate state for trillions of years?

If other intelligent species exist, are they judged by the same standards? Did they fall like humanity? Do they need redemption? Does Christ's work on Earth affect them, or do they have their own salvation history? The cosmos might contain moral dramas we know nothing about.

And from cosmic perspective, human moral significance seems vanishingly small. Our entire species exists for a cosmic eye-blink. Our greatest saints and worst sinners barely register on universal scales. How can choices made by temporary arrangements of atoms on one planet matter eternally?

Perhaps judgment isn't cosmic trial but natural consequence. We become what we choose. Heaven and hell might be the same reality experienced differently based on our orientation—those who learned love experiencing divine presence as joy, those who chose isolation experiencing it as torment. Judgment might be less about punishment and more about truth—finally seeing ourselves as we really are.

Living Toward Heat Death

How should Christians live knowing the universe appears headed toward heat death? Not with denial—ignoring scientific findings undermines credibility and prevents genuine dialogue. Not with despair—current physics doesn't determine ultimate reality. Not with escape into purely "spiritual" salvation that abandons creation God pronounced good.

Instead, we live with hope grounded not in physical extrapolation but in trust that the God who created can recreate. We work for creation's flourishing even if temporary, because present actions have eternal significance in ways we can't calculate. We pursue science to understand creation better, art to celebrate its beauty, ethics to honor its value—not because these prevent heat death but because they participate in meaning that transcends entropy.

This tension—accepting scientific predictions while maintaining eschatological hope—requires spiritual and intellectual maturity. We must hold both truths without resolving their conflict, trusting that resolution exists even if we can't see it. It's like standing at the Red Sea with Pharaoh's army approaching, believing in deliverance while seeing no way forward.

Death and Transformation

Perhaps death—including cosmic death—isn't the opposite of life but its transformation. Stars die to create elements for new worlds. Organisms die to enrich soil for new life. Even black holes evaporate into radiation that might seed future structures. Death seems to be creation's way of making room for new possibilities.

Could heat death itself be transformation rather than termination? Not the end of the story but transition between chapters? The universe achieving maximum entropy might be like a caterpillar entering chrysalis—apparent death preceding unimaginable transformation. The very completion of entropy might create conditions for new creation we can't anticipate.

This doesn't minimize death's reality or pain. The universe will genuinely end as we know it. Everyone we love will die. Everything we've built will crumble. The cosmos itself will go dark. But Christianity has always proclaimed that death doesn't have the last word. Easter follows Good Friday. Resurrection follows crucifixion. New creation follows cosmic death.

The Wager of Hope

Faced with cosmology's prediction of heat death and Christianity's promise of new creation, we must choose how to live. This isn't a choice based on proof but on wager—Pascal's wager extended to cosmic scales.

If physics is right and heat death is final, nothing we do ultimately matters. Our loves, achievements, and sacrifices all end in frozen emptiness. But if Christianity is right and transformation is possible, everything we do has eternal significance. Every act of love defies entropy. Every choice for beauty protests against decay. Every movement toward truth participates in reality that transcends physical description.

The rational wager, even without certainty, is hope rather than despair. Not because hope is proven but because despair is self-defeating. Hope at least allows for possibility. It keeps us open to surprises, responsive to grace, alert to transformation. And if we're wrong, we'll never know it—consciousness will end before it can register its error.

But this hope must be honest. We must acknowledge the genuine challenge heat death poses. We must admit we don't know how God might transform creation. We must recognize that our eschatologies are provisional, metaphorical, pointing toward truth we can't fully grasp.

The universe revealed by modern cosmology often seems hostile to meaning, purpose, and hope. Yet in this universe, consciousness emerged. Love became possible. Beauty came to be appreciated. These realities seem as fundamental as entropy, even if physics can't quantify them. They point toward dimensions of existence that transcend physical description.

Perhaps the deepest eschatological truth is that the God who brought cosmos from chaos, life from non-life, consciousness from matter, can bring new creation from heat death. Not because we understand how but because the pattern of transformation runs through all God's works.

The universe will end. But perhaps it will end like a movement in symphony ends, to be followed by another in a different key. Or end like winter ends in spring—not cessation but transformation. Or end like caterpillar ends to become butterfly—the same identity in unrecognizably different form. Or end in ways we cannot imagine because we've never experienced transformation on such scales.

This is the eschatological wager: that Love is stronger than entropy, that Life transcends death, that the Creator of quantum fields and galaxies has purposes that heat death cannot thwart. We bet our lives on this hope, not because physics confirms it but because the alternative—that consciousness arose by chance only to witness its own inevitable extinction—seems the greater absurdity.

In the face of cosmic death, we choose life. In the face of ultimate entropy, we choose creativity. In the face of heat death, we choose the warmth of faith, hope, and love. And we trust that the God who raised Jesus from the dead can and will raise the cosmos itself to new and unimaginable life.

Prayer in an Expanding Universe

The Absurdity and Necessity of Cosmic Prayer

In a universe of two trillion galaxies, where Earth is less than a speck of dust, where humanity exists for a cosmic eyeblink, prayer seems absurd. I stand in my backyard at night, looking up at stars whose light left before humans existed, and I'm supposed to believe the Mind behind all this cares about my job interview tomorrow? My sick child? My broken heart? The scale differential is incomprehensible—like expecting the ocean to notice a particular molecule, or the galaxy to attend to a specific atom.

Yet humans pray. Across cultures, throughout history, in crisis and celebration, we speak into what seems like void. My grandmother prayed every morning until she died at ninety-three. My physicist colleague prays before lectures. The woman next to me on the plane clutches her rosary during turbulence. Either this is universal delusion—projection of consciousness onto unconscious cosmos—or it reflects something true about reality's nature. Prayer is either the most absurd human activity or the most essential. There's no comfortable middle ground.

When I pray now, informed by cosmology, I feel the absurdity acutely. Here I am, an arrangement of atoms on a tiny planet, sending thoughts into a universe that's 99.9999% hostile vacuum. The nearest star would take light four years to reach. The Andromeda Galaxy, visible to my naked eye, is 2.5 million light-years away. And I'm asking for healing, for guidance, for help with decisions that wouldn't even register as quantum fluctuations on cosmic scales.

But I also feel the necessity. Something in me needs to speak, to reach out, to seek connection with What Is Beyond. It's not rational. It doesn't make scientific sense. But neither does consciousness, or love, or the universe existing at all. Maybe prayer belongs to that category of realities that transcend scientific description while being no less real.

The Problem of Divine Attention

If God is omnipresent throughout an expanding universe that grows larger every moment, does divine attention become increasingly diluted? Are our prayers competing with potential prayers from billions of other worlds? The observable universe contains roughly 10^24 stars. If even a tiny fraction have planets with conscious beings, the prayer traffic must be overwhelming. How can infinite God have intimate relationship with finite creatures when there might be infinite creatures?

The classical answer—that God is infinite and therefore unlimited in attention—feels abstract when I'm on my knees at 3 AM, desperately praying for my mother's cancer treatment. Infinity is a mathematical concept, but I need personal presence. It's one thing to affirm divine omnipresence intellectually, another to believe it existentially while aware of cosmic immensity.

I think the problem is our model of attention as limited resource. We imagine God like an overworked switchboard operator, trying to handle countless calls simultaneously. But consciousness might not work this way. A parent's love doesn't diminish with more children—if anything, it expands. A teacher's knowledge isn't depleted by sharing. Maybe divine consciousness relates to creation through modes that transcend our zero-sum assumptions.

Perhaps God's attention is more like light from the sun—fully present to each point it reaches without being divided or diminished. Or like how the laws of physics are fully present everywhere without being stretched thin. Each sparrow that falls has God's complete attention, not part of it. Each prayer is heard fully, not partially. This is mystery beyond comprehension, but so is consciousness itself.

Prayer and the Block Universe

If relativity is correct and past, present, and future all exist in four-dimensional spacetime, what's the point of petitionary prayer? From God's eternal perspective, my future prayer and its answer (or non-answer) already exist. I'm asking God to change what, from the divine viewpoint, is already fixed in the block universe.

This problem tormented me for years. If the future exists as definitely as the past, prayer seems pointless. It's like asking an author to change chapter 10 when the entire book is already written. We perform prayer theater, acting out lines already scripted, pretending our requests matter when the outcomes are already determined.

Various theological moves try to address this. Some say our prayers don't change God's eternal plan but are part of it. From our temporal perspective, we pray and God responds.

From God's eternal perspective, our prayer and divine response are simultaneous aspects of unified reality. The prayer "causes" the answer from within time even though both exist eternally in God's view.

Others suggest the block universe contains our free choices, including prayers. The future exists but exists partly because of our prayers, not despite them. Our prayers are woven into the fabric of spacetime, part of how the future comes to be what it is. We're not trying to change the tapestry but we are threads within it.

But these explanations feel like clever philosophy that doesn't touch the lived experience of prayer. When I'm praying for my friend's surgery, I'm not trying to be a thread in spacetime's tapestry. I'm pleading for healing. I want God to act, to intervene, to change what seems headed toward tragedy. The block universe makes this kind of prayer seem naive at best, delusional at worst.

When the Universe Doesn't Answer

Let's be honest about what everyone who prays knows but rarely admits: most prayers seem to go unanswered. The cancer isn't healed. The relationship isn't restored. The war doesn't end. The job doesn't come through. The depression doesn't lift. We pray into what feels like emptiness and receive silence in return.

Statistical studies of prayer's effectiveness show mixed results at best. Some studies find small positive effects, others find none, and the effects that are found are usually barely above statistical noise. If God answers prayer, the divine response seems indistinguishable from chance. This is devastating for those of us who want prayer to work like technology—input request, receive result.

The standard explanations feel hollow when you're in the

midst of unanswered prayer. "God knows best" doesn't comfort the mother watching her child suffer. "God won't override free will" doesn't help victims of others' choices. "Prayer changes us, not circumstances" isn't what we want when circumstances are crushing us.

I prayed for my father's healing for two years. I prayed with faith, with doubt, with anger, with resignation. I prayed alone and with others. I prayed traditional prayers and made up my own. He died anyway, slowly and painfully. Either God didn't hear, didn't care, or didn't act. None of those options is comforting.

Yet I still pray. Not because I understand how it works or have evidence that it works, but because something in me needs to pray. It's like breathing—I can hold my breath for a while, but eventually I have to breathe. I can stop praying for a while, but eventually I have to speak into the Mystery, even if the Mystery doesn't speak back in ways I can recognize.

Prayer as Quantum Phenomenon?

When quantum mechanics emerged, some religious thinkers got excited. Here was indeterminacy in nature! Room for divine action! Perhaps prayer works through quantum mechanisms—consciousness affecting probability distributions, entanglement enabling non-local connection, observation collapsing possibilities toward prayed-for outcomes.

I understand the appeal, but this is almost certainly wrong. Quantum effects typically "decohere" in warm, wet, noisy biological systems. By the time you get to anything as large as a cell, quantum superposition has long since collapsed into definite states. The brain is far too warm and noisy to maintain quantum coherence. If quantum effects influence neural activity, it's through randomness, not through conscious control.

Besides, even if consciousness could influence quantum events, this wouldn't explain answered prayer. Nudging a few quantum events in someone's brain wouldn't cure cancer or end war or bring rain. The scale is wrong. It's like trying to steer a ship by blowing on individual molecules of water.

Yet there might be a subtler truth here. If consciousness emerges from quantum processes—and we don't understand how—then prayer involves consciousness, which involves quantum mechanics at some deep level. Not that prayer is quantum phenomenon, but that both prayer and quantum mechanics might point toward deeper realities neither fully captures. Both involve the relationship between observer and observed, the nature of information, the basis of physical law.

Maybe prayer doesn't work through quantum mechanics but shares with quantum mechanics a participation in layers of reality we don't understand. Both remind us that the universe is stranger than common sense suggests, that consciousness and cosmos are connected in ways we can't fully map.

Contemplation and Cosmic Awareness

Contemplative prayer—wordless attention to divine presence—takes on new dimensions in cosmic context. When we quiet mental chatter and become present to What Is, we might touch something that transcends local space and time. Mystics across traditions report experiences of cosmic consciousness—awareness expanding beyond individual boundaries to encompass vast scales.

I've had moments in meditation where the boundary between self and cosmos seems to dissolve. Not frequently, not on command, but occasionally—usually when I'm not trying—awareness expands beyond my small self. I become conscious of existing in a vast universe, made of stardust, product of billions of years of cosmic evolution. The atoms in my body were forged in stellar cores. The water I drink is older than Earth. I am the universe becoming aware of itself.

Are these moments delusion—oxygen deprivation or temporal lobe activity creating illusions of transcendence? Or glimpses of reality from enlarged perspective? I can't prove they're valid, but they feel more real than ordinary consciousness, not less. They leave me changed in ways that suggest encounter rather than projection.

Modern cosmology provides new icons for contemplation. The Hubble deep field image, showing thousands of galaxies in what appears to be empty space. The cosmic web, revealing the large-scale structure of universe. The Pale Blue Dot, Earth photographed from beyond Neptune, a mote of dust suspended in sunbeam. These images can trigger contemplative awareness if we let them, opening us to scales beyond ordinary experience.

Intercessory Prayer Across Cosmic Distance

When we pray for others, especially those far away, we participate in something that transcends ordinary causation. I pray for my friend in Japan, for refugees in Syria, for astronauts on the space station. These prayers can't be traveling through physical space—they would take years to reach even the nearest star. If prayer works at all, it works through dimensions that transcend space.

This isn't magical thinking but recognition that consciousness and relationship might involve aspects of reality physics hasn't mapped. We know about quantum entanglement—particles connected in ways that transcend space. We know about fields that pervade all space—electromagnetic, gravitational, quantum. Maybe prayer operates through dimensions or fields we haven't discovered, ways consciousness connects that transcend physical separation.

Or maybe prayer doesn't travel at all because God is already present everywhere. We're not sending messages across cosmic distances but participating in divine presence that pervades all points of spacetime. The distance between me and the person I'm praying for is real from our perspective but irrelevant from God's. We're all equally in divine presence, equally close to the Heart of reality.

This makes intercessory prayer both easier and harder. Easier because distance doesn't matter. Harder because we can't control outcomes through prayer technology. We can only hold others in awareness, in love, in hope, trusting that this matters in ways we can't measure or prove.

Personal Prayer When You're Cosmically Insignificant

The cosmos' vastness can make personal concerns seem trivial to the point of absurdity. Why pray about employment when stars are exploding? Why mention relationship struggles when galaxies are colliding? Why bring my small anxieties to the Creator of two trillion galaxies?

Yet the incarnation suggests God values the particular, the local, the personal. The Creator of quantum fields became a particular person with specific concerns—hunger, fatigue, grief over a friend's death. If God can be incarnate, God can care about incarnate beings' needs. The same God who sustains galaxies notices sparrows falling.

The challenge is holding both truths simultaneously. We are cosmically insignificant—spatially negligible, temporally brief, easily eliminated by countless cosmic hazards. And we are eternally significant—conscious beings capable of love, beauty, moral choice, and relationship with divine. Our problems are trivial on universe scales and important to the God who numbers hairs and knows our names.

When I pray now, I try to hold this double awareness. "God of two trillion galaxies, I'm struggling with this decision..." It sounds absurd, but so does existence itself. If the universe can produce consciousness capable of comprehending the universe, maybe that consciousness matters more than its physical insignificance suggests.

Prayer as Participation in Divine Consciousness

Perhaps prayer is less about communication than participation. Not sending messages to distant God but sharing in divine awareness that pervades all reality. When we pray, we're not informing God—who already knows—but becoming conscious of divine consciousness of our situation.

This changes prayer from transaction to transformation. I'm not trying to get God to do something but trying to align myself with what God is already doing. Not convincing God to care but becoming aware of divine care that already exists. Not bridging distance but recognizing presence.

From this perspective, unanswered prayer isn't divine absence but our inability to perceive divine presence and action. The healing we seek might come through medical science, human compassion, or inner strength we didn't know we had. The answer might be "no" because what we request would harm us or others in ways we can't see. Or the answer might be something we don't recognize as answer because it doesn't match our expectations.

This doesn't solve the problem of suffering or explain why prayers for obviously good things—end to war, healing of children, relief from oppression—seem to go unanswered. But it does suggest that prayer might be more about participation than petition, more about relationship than results.

The Practice of Cosmic Prayer

How might cosmic awareness transform prayer practice? Instead of immediately launching into requests, I now often begin by acknowledging scale. I think about the atoms in my body, forged in stars. The billions of years that led to this moment. The vast spaces surrounding our tiny world. This isn't to diminish my concerns but to contextualize them.

Sometimes I practice what I call "scale meditation"—consciously moving through scales from quantum to cosmic. Starting with subatomic particles, moving through atoms, molecules, cells, my body, Earth, solar system, galaxy, universe. Then returning, bringing cosmic perspective to personal concerns. It's humbling and oddly comforting.

I pray with awareness of deep time. This moment contains billions of years of history. The carbon in my body, the oxygen I breathe, the ground I kneel on—all products of cosmic evolution. My prayer is the universe becoming conscious of itself and reaching toward its source. I'm not separate from cosmos but cosmos in the act of prayer.

And I pray with others across space and time. When I recite ancient prayers, I join voices across millennia. When I pray liturgically, I join communities around the world. We're nodes in a network of consciousness, each praying from our unique location in spacetime but participating in something that transcends spacetime.

Honest Assessment: Prayer's Uncertain Efficacy

Does prayer work? After a lifetime of praying and decades of studying both physics and theology, my honest answer is: I don't know. Effects, if they exist, are subtle enough to be debated. Prayer doesn't work like technology or even like medicine. Its efficacy, if real, operates through channels we can't measure or understand.

This uncertainty is frustrating but perhaps appropriate. If prayer's effects were obvious and repeatable, it would be technology, not relationship. If we could control outcomes through proper prayer technique, we'd be manipulating God rather than relating to God. The ambiguity preserves both human freedom and divine mystery.

We pray not because we're certain it works but because we're human. Prayer expresses our nature as conscious, meaning-seeking, relationship-oriented beings. Even if prayer doesn't affect external reality (though it might), it affects us. It shapes our consciousness, orients our attention, opens us to possibilities beyond our control.

The Persistence of Prayer

Despite cosmic scales that dwarf us, despite physics that suggests a block universe, despite unanswered prayers and uncertain efficacy, humans persist in praying. This persistence is either testimony to delusion's power or evidence of something true about reality's nature.

Modern cosmology doesn't resolve prayer's mystery but provides new context for it. We pray as products of 13.8 billion years of cosmic evolution, as arrangements of stardust that developed consciousness, as the universe becoming aware of itself and reaching toward its source. Our prayers might be absorbed into cosmic void or heard by cosmic Consciousness. They might change nothing or participate in everything.

When I pray now, I hold all these possibilities simultaneously. I speak words that might echo unheard in empty space or might be more fundamental than space itself. I ask for outcomes I know might be fixed in spacetime's block or might be genuinely open to divine action. I seek relationship with Mystery that might be projection or might be recognition of reality's deepest truth.

In the expanding universe, where space itself grows larger every moment, where galaxies race apart toward isolation, prayer is act of resistance. It insists on connection despite distance, meaning despite vastness, relationship despite scale. Whether this resistance is futile or fundamental remains the question each person must answer in the solitude of their own cosmic prayer.

But perhaps the question is the answer. Perhaps the fact that we pray—that atoms arranged in certain patterns seek communion with Whatever Arranged Them—is itself the evidence we seek. The universe prays through us. And in that cosmic prayer, absurd and necessary, we touch something more real than stars.

Ethics in Cosmic Perspective

The Moral Weight of Stardust

We are arrangements of atoms forged in dying stars, products of 13.8 billion years of cosmic evolution, possibly the only moral agents in the observable universe. This morning, as I made the small decision to return extra change at the coffee shop, I was acutely aware of the absurdity—a collection of particles that once burned in stellar cores choosing honesty over gain, as if it matters to a universe that doesn't know we exist.

Yet maybe that's precisely why it matters. If we are how the universe becomes capable of moral choice, then our ethics carry cosmic significance. We're not just choosing for ourselves but as the universe's moral agents, its way of developing conscience. The atoms in my hand that returned those few dollars have traveled billions of years and light-years to arrive at this moment of choice. Either that journey culminates in meaninglessness, or moral choice is what gives it meaning.

The traditional religious answer—we have duties to our Creator—remains but gains cosmic scope when we grasp what creation actually is. The secular answer—we create our own meaning—becomes more poignant when set against cosmic indifference. Either way, ethics can't be separated from cosmology. What we are affects what we ought to do. And what we are is spectacular and unlikely: matter that has developed the capacity to choose between right and wrong.

The Problem of Cosmic Indifference

The universe revealed by modern cosmology seems spectacularly indifferent to moral concerns. This morning's news reports an earthquake that killed thousands. Tonight, a supernova is sterilizing any planets within dozens of light-years. Stars explode without regard for the civilizations they might destroy. Black holes devour indiscriminately. Asteroids impact worlds randomly, ending billions of years of evolution in moments. The same physics that enables life enables cancer. The cosmos appears amoral at best, actively hostile at worst.

Standing in a children's cancer ward, watching parents hold dying children, the cosmic indifference feels unbearable. These children are made of stardust, products of billions of years of cosmic evolution, and the universe that made them is killing them through random genetic mutations. Where is justice in quantum randomness? Where is purpose in purposeless suffering?

Against this backdrop, human moral concerns seem like projections onto indifferent reality. Why care about justice in a universe trending toward heat death? Why preserve life that cosmos will eventually extinguish? Why act ethically when physics is fundamentally amoral? The universe offers no moral guidance. Stars don't teach justice. Galaxies don't model compassion. Quantum fields don't demonstrate dignity.

Some find this liberating—without cosmic moral order, we're free to create our own values, unconstrained by divine commands or natural law. Others find it devastating—ethics becomes arbitrary preference with no grounding in reality, no foundation deeper than human consensus. I find myself caught between these responses, unable to accept either fully. The universe seems indifferent to ethics, yet here I am, incapable of indifference, feeling moral obligations I didn't choose and can't escape.

Evolution and the Roots of Morality

We evolved through processes involving competition, predation, and death. Our moral intuitions emerged from evolutionary pressures that had nothing to do with truth or goodness and everything to do with survival and reproduction. Altruism toward kin helped our genes propagate. Reciprocity with neighbors provided mutual benefit. Hostility toward outsiders protected resources. These are the origins of what we call morality.

This evolutionary history creates uncomfortable questions. When I feel compassion for strangers, am I experiencing genuine moral insight or evolutionary misfiring—my kin-protection instincts over-generalizing? When I feel righteous anger at injustice, is this moral truth or tribal psychology—my group-defense mechanisms activating? Are ethics anything more than evolutionary strategies that got elaborated into philosophical systems?

Our evolutionary baggage complicates every ethical decision. We evolved in small groups, generating in-group loyalty and out-group suspicion. This poorly equips us for global ethics, much less cosmic ethics. We're trying to apply moral intuitions developed for tribes to nations, species, and potentially other forms of consciousness. It's like using a compass designed for Earth to navigate galactic space.

We evolved for short-term thinking—immediate threats and opportunities mattered more than long-term consequences. This makes addressing climate change or planning for cosmic futures almost impossibly difficult. Our brains resist thinking in geological time, much less cosmic time. We feel tomorrow's problems intensely but can barely grasp problems a century away.

Violence served evolutionary purposes. Aggression helped our ancestors compete for resources and mates. Our capacity for violence is as natural as our capacity for love—both written into our genes by millions of years of selection. When I suppress violent impulses, am I overcoming nature or expressing it? When I choose peace over conflict, is this moral progress or evolutionary strategy?

Yet evolution also produced capacities that transcend evolutionary logic. We appreciate beauty that serves no survival value—abstract art, pure mathematics, sunsets we watch alone. We develop moral intuitions that sometimes oppose genetic interests—celibacy, adoption of unrelated children, sacrifice for strangers. We seek truth even when comforting lies would help us survive.

Perhaps evolution equipped us not just with specific moral intuitions but with moral capacity itself—the ability to reflect on our intuitions, evaluate them, sometimes reject them. We're not slaves to evolutionary programming but conscious beings capable of transcending our origins. The same process that gave us tribalism also gave us the ability to recognize tribalism as a problem.

Creation Care or Cosmic Futility?

If the universe faces heat death, if Earth will be consumed by the sun in a billion years, if everything ends in frozen emptiness, why preserve the environment? Why protect species that will eventually go extinct? Why maintain what cosmos will destroy?

I stand in an old-growth forest, trees that have lived for centuries, ecosystems of incredible complexity and beauty, and I know it's all temporary. These trees will die. This forest will vanish. This planet will be incinerated. On cosmic timescales, the difference between clear-cutting this forest today and preserving it for another million years is negligible. So why does destroying it feel wrong?

Several responses compete in my mind. Perhaps creation has intrinsic value independent of duration. A flower is beautiful even though it fades. A species is precious even though it's temporary. A forest matters even though it won't last forever. Value isn't determined by permanence.

Or perhaps we're accountable for our actions regardless of ultimate outcomes. Ethics isn't about final results but immediate choices. I'm responsible for what I do with the time and place I'm given, not for cosmic outcomes beyond my control. My ethics are local even if the universe is infinite.

But honest assessment acknowledges the tension. Working to preserve what will be destroyed requires either hope beyond evidence or commitment beyond hope. When I pick up litter, plant trees, or work for environmental protection, I'm acting as if the future matters while knowing the future ends in heat death. This is either noble defiance or pathetic delusion.

Technology and Cosmic Power

Human technology approaches cosmic significance. We can alter planetary climate through industrial emissions. We can edit the code of life through genetic engineering. We're creating artificial intelligence that might exceed human cognition. We might soon be able to spread Earth life throughout space, terraform other worlds, or trigger our own extinction. We're making decisions with million-year consequences using wisdom developed for hunter-gatherer societies.

The ethical challenges are unprecedented. Should we use genetic engineering to eliminate diseases? What about enhancing human capabilities? Who decides what constitutes improvement? We have the power to direct evolution but not the wisdom to know where it should go.

Should we create artificial general intelligence? If consciousness is substrate-independent, we might create minds that suffer or flourish based on our design choices. We'll be responsible for new forms of consciousness we don't understand. Can an AI have rights? Can it be wronged? If we create conscious machines and then shut them down, is that murder?

Should we terraform Mars? Do we have the right to transform another world, even a lifeless one? What if there's microbial life we haven't detected? Does that life have moral status? Is it better to spread Earth life through the cosmos or leave other worlds pristine?

These aren't distant future scenarios but near-term decisions. Within decades, we'll face choices that affect cosmic futures. We're like children who've found the keys to construction equipment—capable of tremendous creation or destruction without understanding the full consequences of either.

The Ethics of Extinction

On cosmic timescales, extinction is normal. 99.9% of all species that ever lived are extinct. Mass extinctions punctuate evolutionary history—the Permian extinction eliminated 90% of species, the Cretaceous extinction ended the dinosaurs. Eventually, all current species will go extinct, including humans. How does this affect our moral obligations?

If extinction is inevitable, why fight to save endangered species? The polar bear will go extinct eventually whether we save it now or not. If humanity will eventually end, why ensure our survival? Why not enjoy what time we have without worrying about future generations who are doomed anyway?

Yet the inevitability of extinction might make life more precious, not less. Each species is unique, irreplaceable, a one-time experiment in consciousness and complexity. When a species goes extinct, its particular way of being in the world vanishes forever. The universe becomes less diverse, less interesting, less itself.

But this raises hard questions about intervention. Should we preserve species that evolution would eliminate? Natural selection is cruel but creative—extinction makes room for new forms. By preventing extinctions, are we stopping evolution? Should we use technology to preserve genetic information of extinct species for potential future resurrection? Are we obligated to spread life beyond Earth to prevent total extinction from planetary catastrophe?

Justice Across Deep Time

Our actions affect future generations across potentially cosmic timescales. Climate change will impact thousands of years. Nuclear waste remains hazardous for millions. Genetic modifications could affect all future humans. Space ventures might have consequences lasting billions of years. How do we weigh present benefits against future costs when the future extends toward infinity?

Traditional ethics focuses on contemporary justice—fair distribution among current people. But cosmic ethics must consider vast temporal scales. Every decision ripples forward through time. The carbon I emit today affects climate for centuries. The species we eliminate can never be recovered. The AI we create might outlive our species.

How do we value future people who don't yet exist? How many future generations matter morally? If we consider all potential future humans, their numbers dwarf current humanity. Should we sacrifice present welfare for potentially infinite future welfare? But if we don't consider them, we're treating the future as morally irrelevant.

No ethical framework adequately addresses such scales. We're making cosmic decisions with tribal morality, affecting deep time with short-term thinking. It's like trying to perform surgery with stone tools—we have the power but not the precision.

Moral Relativism in Space

If intelligent life exists elsewhere, evolved through different processes under different conditions, would they share our moral intuitions? Is ethics universal like mathematics—would aliens discover the same moral truths we have? Or is ethics evolutionary accident, varying with biology and history?

Mathematics seems universal. Aliens would discover the same prime numbers, the same physical laws. But would they recognize dignity, justice, compassion? Or would their ethics be as alien as their biology? A hive mind might not value individuality. Silicon-based life might not understand suffering. Beings with different reproductive strategies might have completely different concepts of family, loyalty, and care.

This isn't abstract speculation. If we encounter alien life or create artificial consciousness, we'll need ethical frameworks that transcend human evolutionary history. But we have no idea what universal ethics might look like. Our deepest moral convictions might be biological prejudices. What seems obviously right to us might be incomprehensible to minds evolved under different conditions.

Even on Earth, moral intuitions vary dramatically across cultures. Expanding to cosmic scales multiplies the challenge infinitely. We might encounter beings whose ethics we find abhorrent or incomprehensible. How do we adjudicate between incompatible moral systems without shared foundation?

The Virtue of Cosmic Humility

Modern cosmology teaches humility. We don't understand 95% of the universe. We can't predict long-term consequences of our actions. We don't know if we're alone. We might be wrong about fundamental aspects of reality. This cosmic humility should shape our ethics.

Humility doesn't mean paralysis but appropriate caution. When actions might cause irreversible harm on cosmic scales, uncertainty shouldn't delay prevention. We should prefer reversible choices over permanent ones, preserve options rather than optimize for single values, test carefully before implementing broadly.

But humility also means recognizing our limitations. We're not cosmic engineers capable of managing universal outcomes. We're local agents with limited knowledge and power. Our ethics must be appropriate to our scale—doing what good we can where we are, without pretending to control cosmic futures.

Love in the Cold Cosmos

In a universe trending toward heat death, where entropy always increases, where most of space is cold vacuum hostile to life, love seems like cosmic anomaly. The universe doesn't love. Stars don't care. Galaxies collide without compassion. Yet here we are, collections of atoms capable of love, products of blind evolution who developed sight, outcomes of indifferent processes who care intensely.

When I hold my daughter, both of us made of stardust, both of us temporary arrangements of matter in an indifferent cosmos, the love I feel seems simultaneously insignificant and ultimate. Insignificant because we're specks in infinity. Ultimate because this might be the only love in the observable universe—conscious beings choosing connection in a cosmos that doesn't require it.

If love is accidental byproduct of evolution, it has no cosmic significance. But if consciousness and care are what cosmos is developing toward, then love might be as fundamental as physics—perhaps more so. Maybe the equations describe the stage, but love is the play. Maybe matter is the medium, but meaning is the message.

Christian ethics grounds itself in love—God's love for creation, human love for God and neighbor. Cosmic perspective doesn't diminish this but extends it. Love your neighbor must now include future generations, other species, potential artificial minds, maybe alien civilizations. But can love stretch so far without breaking? Can we genuinely care about beings we'll never meet on timescales we can't comprehend?

Living Ethically in an Amoral Universe

How do we maintain moral seriousness in a cosmos that seems amoral? The universe offers no validation for ethical choices. Good people suffer, evil people prosper, and the cosmos continues indifferent to both. Stars shine on just and unjust alike. Entropy increases regardless of our moral choices.

Yet we make moral choices anyway, as if they matter ultimately. When I choose honesty over deception, kindness over cruelty, justice over oppression, I'm asserting meaning the universe doesn't confirm. This is either profound delusion or profound truth—maybe the highest truth, that conscious beings can create moral reality in amoral cosmos.

We act ethically not because the universe rewards virtue but because we're the kind of beings who experience moral obligation. Ethics expresses our nature as conscious agents, regardless of cosmic response. We're the universe's way of developing conscience. Through us, cosmos becomes capable of moral choice, ethical reflection, and chosen goodness.

The Burden and Gift of Moral Agency

We are possibly the only moral agents in the observable universe. This is simultaneously crushing burden and extraordinary gift. We bear responsibility for choices affecting cosmic futures with wisdom barely adequate for local presents. Every decision ripples outward through space and forward through time in ways we can't fully trace.

Modern cosmology doesn't provide ethical answers but reframes ethical questions. Traditional concerns—justice, compassion, dignity—remain but require translation to cosmic scales. We need ethics for deep time, vast spaces, and unprecedented powers. We need wisdom for decisions affecting billions of years and potentially billions of worlds.

The universe offers no moral guidance. If values exist, they exist because consciousness brings them into being. We are how cosmos develops conscience. This is terrifying responsibility and amazing opportunity. In us, the universe becomes capable of choosing good over evil, love over indifference, meaning over meaninglessness.

We will make mistakes—perhaps cosmic-scale mistakes. Our ethics, evolved for tribal life, poorly equip us for cosmic decisions. Yet we must decide and act, creating ethics adequate to our cosmic moment. In the cold, expanding universe, every act of love is rebellion against entropy. Every choice for justice defies cosmic indifference. Every preservation of beauty protests against heat death.

These rebellions might be futile—brief flickers of meaning in infinite darkness. Or they might be fundamental—the universe discovering its purpose through creatures capable of choosing purpose. We don't know which, but we must live as if our choices matter infinitely. Because if we are the only moral agents in the cosmos, then our choices are the only ones that do matter.

We are stardust contemplating ethics, matter pondering morality, the universe developing conscience. This is either absurd accident or sacred calling. Probably both. In the tension between cosmic indifference and moral conviction, we create lives worth living in a universe that won't last forever. And perhaps, in that creation, we touch something that will.

Scientists, Faith, and Honest Struggles

Beyond the Success Stories

Most books about science and faith feature scientists who've achieved perfect integration—eloquent testimonies of harmony between research and religion, stories polished until they gleam like Nobel medals. These narratives inspire but can also mislead, suggesting the tension dissolves if you're just smart enough, faithful enough, or sophisticated enough in your thinking.

I want to tell a different story. Or rather, I want to let others tell their stories—messier stories, ongoing struggles, unresolved tensions that persist despite brilliant minds and deep faith. These are the conversations that happen in faculty offices after the conferences end, in late-night discussions when defenses drop, in the honest moments when scientists who are believers admit they don't have it all figured out.

Over the years, I've collected these stories, changed names

and some details to protect privacy, but preserved the essential honesty. These aren't testimonies of triumph but of persistence, not of resolution but of resilience. They're stories of people who continue both scientific research and religious practice not because they've solved all contradictions but because abandoning either would diminish their humanity.

Dr. Sarah Chen: The Cosmologist's Paradox

Sarah studies the cosmic microwave background radiation, those ancient whispers from the Big Bang that still echo throughout space. Her office walls are covered with false-color maps showing temperature fluctuations of a hundred-thousandth of a degree—variations that seeded all structure in the universe, the primordial pattern from which galaxies would eventually emerge.

"People assume I see God in these patterns," she tells me over coffee in the faculty lounge. "They expect me to point to the CMB and say, 'Look, divine fingerprints!' But honestly? Most days I see exactly what my atheist colleagues see—quantum fluctuations in a hot, dense plasma, stretched by cosmic expansion, nothing more."

She pauses, stirring her coffee thoughtfully. "But then Sunday comes, and I'm teaching seven-year-olds in Sunday school. Last week, little Emma asked if God lives in space. How do you answer that? I stumbled through something about God being everywhere, but I was thinking about the vast emptiness between galaxies, the hostile vacuum that would kill her in seconds. I was thinking about dark energy pushing everything apart, about heat death, about the universe becoming infinitely large and infinitely empty."

The tension is palpable in her voice. "When I model the universe's expansion, calculating how dark energy will eventually tear everything apart, it's hard to sing 'This Is My Father's World' without feeling like a fraud. When I realize that 99.999% of the universe would instantly kill any life we know, claiming God designed it for us seems... presumptuous at best."

Her atheist colleagues assume she compartmentalizes—scientist during the week, believer on Sunday, never the twain shall meet. "But it's not that clean," she insists. "My faith and science interpenetrate in ways that sometimes harmonize, sometimes clash, often just create this uncomfortable dissonance I've learned to live with."

She shows me her latest paper on cosmic inflation, equations describing the universe's first fraction of a second. "Fine-tuning genuinely amazes me. The cosmological constant's precision seems to point beyond itself, to whisper of intention. But I also understand anthropic reasoning, selection effects, the possibility of multiverses. I can't claim fine-tuning proves God without acknowledging other explanations. So I hold it lightly, this hint that might be evidence or might be projection."

Prayer is particularly challenging for her. "I pray for my mother's cancer treatment while knowing prayer studies show mixed results at best. I thank God for the beauty of spiral galaxies while knowing they formed through gravitational instabilities, not aesthetic design. Sometimes prayer feels like talking to myself in an empty universe. Sometimes it feels like participating in something deeper than physics can measure. Most times, I honestly can't tell which it is."

Dr. James Okonkwo: Evolution's Uncomfortable Truths

James is a Catholic evolutionary biologist, and yes, he sees the irony. His research focuses on parasitic relationships, the exquisite mechanisms by which organisms exploit others. His laboratory refrigerators contain samples that would horrify most people—parasites that zombify their hosts, wasps that use living caterpillars as nurseries for their young, fungi that control ant behavior before erupting from their heads.

"I study the mechanics that many of my fellow believers reject or sanitize," he says, showing me electron microscope images of a parasite that systematically replaces a fish's tongue. "I trace genetic relationships that unambiguously show humans and chimps share common ancestors. I document natural selection's power to create complexity without design, purpose, or mercy."

The evolution-creation tension isn't easily dismissed by saying "God used evolution." Evolution involves massive suffering, extinction, and waste. "Parasites that eat children's eyes from inside evolved through the same process that produced human compassion," he points out. "If God guided evolution, divinity chose a brutal method. When I study parasitic wasps that paralyze caterpillars so their larvae can eat them alive from inside, keeping them fresh, I don't see divine wisdom. I see horror that natural selection explains perfectly without invoking purpose."

Yet he remains Catholic. Not because he's solved the problem of natural evil but because the alternative seems even more problematic. "The fact that evolution produced beings capable of moral revulsion at evolution's cruelty suggests something beyond evolution. We're the only species that looks at nature and says, 'This is wrong.' That moral sense, which contradicts our evolutionary interests, points to something transcendent."

Mass presents its own challenges. "The Eucharist proclaims matter's capacity for divine presence. But I know bread is just carbohydrates and proteins, wine is just ethanol and water. Transubstantiation makes no scientific sense." He smiles wryly. "My Protestant friends say I should just accept it symbolically. But that's too easy. The scandal of the Real Presence parallels the scandal of the Incarnation—infinite in finite, creator in creation. If I'm going to believe absurdity, I'll believe it fully."

Research sometimes strengthens his faith. "When I discover genetic networks of extraordinary elegance, witness evolutionary innovations that seem to anticipate future needs, observe symbioses of amazing complexity—these evoke wonder that feels worshipful. Evolution might be blind, but it creates beauty that transcends function."

But research also challenges faith constantly. "Most mutations are harmful. Most species go extinct. Evolution has no goal, no purpose, no care for suffering. If this is God's method, what does that say about God? I live in this tension. During the day, I'm a methodological naturalist, explaining life through purely natural processes. During Mass, I'm a mystical supernaturalist, participating in miracle. The contradiction would be unbearable if I thought either perspective complete."

Dr. Maria Rodriguez: Quantum Faith

Maria's journey is different from the others. She grew up in a fundamentalist church that taught God controlled everything, that nothing happened outside divine will. "I believed in a God who determined every detail, whose plan was precise and knowable. Then I discovered quantum mechanics."

She laughs, but there's pain in it. "Quantum mechanics broke my fundamentalist faith. Reality is fundamentally probabilistic. God doesn't determine where the electron will be, only probability distributions. If God doesn't control quantum events, does God control anything? If reality is genuinely random at base, where is providence?"

She went through what she calls her "atheist phase," convinced physics had disproven her childhood God. "But quantum mechanics also prevented me from settling into materialism.

Reality is too weird. Entanglement, superposition, the role of observation—these aren't properties of classical matter. Consciousness seems involved in ways we don't understand. Information appears more fundamental than substance."

She's developed what she calls "quantum faith"—comfort with uncertainty, acceptance of paradox, peace with probability. "I don't believe in the deterministic God of my childhood, but I believe in Something that transcends current physics. My prayer has changed completely. I don't petition for specific outcomes but align myself with possibility fields. I don't expect intervention but participation."

Her evangelical family thinks she's lost faith. Her physicist colleagues think she's clinging to religious residue. "Maybe both are right. Or maybe I'm finding faith that honors both quantum reality and spiritual experience. The hardest part is community. I don't fit in churches that insist on classical theism. I don't fit in physics departments that assume materialism. I'm theologically homeless, scientifically suspect."

Yet occasionally she glimpses integration. "When equations of breathtaking beauty predict bizarre realities that experiments confirm. When quantum mechanics suggests reality is relational rather than substantial. When consciousness appears fundamental rather than emergent. These moments hint at truth neither physics nor theology alone captures."

Common Threads in the Struggle

Listening to these stories and dozens like them, patterns emerge. These scientists share common struggles that transcend denominational differences or scientific specialties.

The cognitive dissonance is real and persistent. Holding scientific and religious worldviews simultaneously creates mental tension that never fully resolves. They've learned to live with it rather than eliminate it, like musicians who've learned to hear dissonance as part of the composition rather than mistake to be corrected.

Social isolation marks their experience. In academic settings, admitting to religious faith can damage credibility, limit collaboration, affect tenure decisions. In religious communities, accepting evolution or questioning traditional interpretations can lead to suspicion, exclusion, even accusations of heresy. They often feel marginalized in both communities—too religious for academia, too scientific for church.

Prayer remains meaningful but mysterious. They continue to pray despite uncertainty about mechanism or efficacy. Prayer becomes less about getting results and more about maintaining connection, less about changing God's mind and more about changing their own. Yet the ambiguity troubles them—is prayer conversation or monologue, connection or projection?

The problem of natural evil refuses resolution. Cosmic violence, evolutionary cruelty, the universe's apparent indifference to suffering—these challenge belief in benevolent creation. Traditional theodicies feel inadequate when you spend your days studying supernovae that sterilize space or parasites that torture their hosts. Faith persists despite rather than because of evidence.

Integration exhaustion is real. Constantly translating between scientific and religious languages, always explaining to one community why the other isn't crazy, perpetually defending choices that satisfy neither side—it's intellectually and emotionally draining. Some days they want to choose one world and abandon the other, just for the relief of simplicity.

Career costs accumulate. Being openly religious has closed doors, limited opportunities, created awkwardness at conferences. Being scientifically literate has complicated church relationships, eliminated teaching opportunities, created distance from faith communities. They pay a price for refusing to choose sides.

Yet they persist. Not because they've found answers but because both science and faith seem essential to their humanity. Abandoning either would be a kind of death, a voluntary lobotomy, a chosen blindness to dimensions of reality that call to them equally.

The Gift of Honest Struggle

These scientists offer something more valuable than triumphant testimonies: honest struggle. They model what it looks like to live with integrity when integrity means acknowledging rather than resolving tensions. Their stories suggest the science-faith dialogue is more difficult than apologists admit but more possible than skeptics claim.

They've taught me that mature faith in the age of science might look less like certainty and more like commitment, less like answers and more like engagement, less like resolution and more like resilience. They continue their journeys not with confidence of arriving but with determination to keep walking, trusting that truth—scientific and spiritual—is worth pursuing even when the path is unclear.

Their struggles are gifts to those of us who share them. In their honesty, we find permission for our own doubts. In their persistence, we find encouragement for our own journey. In their refusal to simplify, we find validation for our own complexity.

The conversation between science and faith continues in their lives, not as abstract theological exercise but as daily negotiation, lived tension, embodied paradox. They are living laboratories where cosmology and Christianity collide, sometimes producing light, sometimes heat, always generating energy that drives them forward into mystery.

Their witness is not that the tension resolves but that it can be creative, not that questions find answers but that questions worth asking are worth living, not that science and faith harmonize but that the dissonance might itself be sacred—the sound of finite minds engaging infinite reality, of hearts large enough to hold seemingly contradictory truths, of souls brave enough to say "I don't know" while continuing to seek.

Resources for Honest Exploration

A Different Kind of Resource Guide

Most resource chapters read like bibliographies with commentary, neat lists of books and websites sorted by category, each with a brief description of why you should read it. I want to do something different here. Instead of prescribing a curriculum, I want to map the territory of exploration, including resources that will challenge as well as confirm, complicate as well as clarify.

The goal isn't to protect faith from science or science from faith, but to equip you for honest engagement with both. Some resources I'll mention will strengthen faith; others will challenge it to the breaking point. Some will deepen scientific understanding; others will reveal its limits and assumptions. All contribute to the conversation, even—perhaps especially—the ones that make us uncomfortable.

Reading as Pilgrimage

Books have been my companions on this journey between cosmology and faith, and I want to share not just titles but experiences, not just what to read but how these books changed me. Let me walk you through my library, pulling volumes from shelves, remembering where I was when I first encountered them, how they shaped or shattered my thinking.

Sean Carroll's "The Big Picture" sits next to my desk, bristling with sticky notes and marginalia. Carroll is an atheist physicist who writes with clarity and honesty about a naturalistic worldview. No God required or implied. When I first read it, I wanted to argue with every page. Now I return to it regularly, not because I agree but because Carroll forces me to confront the coherence of materialism. He doesn't strawman religion; he simply shows how much can be explained without it. Reading Carroll is like training at altitude—difficult but strengthening.

On the same shelf, Carl Sagan's "Cosmos" leans against Francis Collins's "The Language of God." The juxtaposition is intentional. Sagan's prose about cosmic grandeur, written from a secular perspective, paradoxically evokes worship in me. His wonder at the universe, despite his atheism, seems to point beyond itself. Collins, the geneticist who led the Human Genome Project, writes as a Christian scientist, but sometimes his integration feels too neat, the rough edges sanded smooth. I need both books—Sagan's honest wonder without God, Collins's faith seeking understanding.

The middle shelf holds the complications. William James's "The Varieties of Religious Experience" transformed how I think about faith itself. James, writing over a century ago, treats religion as a human phenomenon worthy of empirical study. He makes faith seem both universal and potentially pathological, both humanity's highest achievement and perhaps its most persistent delusion. After reading James, I could never again assume my religious experiences were self-authenticating.

Pierre Teilhard de Chardin's "The Phenomenon of Man" occupies a strange place in my library and my thinking. This Jesuit paleontologist attempted such radical integration of evolution and Christianity that he satisfied neither scientists nor theologians. His vision of cosmic evolution directed toward the Omega Point (Christ) is either profound or heretical, perhaps both. I don't agree with Teilhard, but he showed me how far integration could be pushed, and perhaps where it breaks.

Harold Kushner's "When Bad Things Happen to Good People" isn't about cosmology, but it belongs in this discussion. Written after his son died from progeria, Kushner challenges traditional theodicy with raw honesty. His conclusion—that God is not omnipotent—scandalized many believers. But for those of us wrestling with cosmic indifference and evolutionary cruelty, Kushner's honesty about divine limitation opens important conversations.

Organizations as Communities of Exploration

The logos of various organizations decorate my conference folders, each representing a different approach to science-faith dialogue. BioLogos, founded by Francis Collins, promotes evolutionary creation through excellent resources. Their website helped me understand how Christians could accept evolution without abandoning faith. But sometimes they minimize genuine tensions, smoothing over difficulties that deserve acknowledgment.

The Faraday Institute at Cambridge takes a more academic approach, hosting conferences where scientists and theologians engage in serious dialogue. I attended one of their summer courses—intense days of lectures and discussions that raised more questions than they answered. That's their strength:

they're better at complexifying than simplifying.

On the other end of the spectrum, organizations like Reasons to Believe attempt to reconcile science with more conservative theology. While I disagree with their old-earth creationism, engaging their arguments helped me understand why some Christians interpret scientific evidence differently. Their challenges to mainstream science sometimes highlight genuine puzzles, even if their solutions don't convince.

The National Center for Science Education defends evolution education in public schools, opposing creationism and intelligent design. Reading their materials helped me understand why scientists find religious anti-evolution arguments so frustrating. Their critiques are worth engaging, even when they sting.

The Templeton Foundation funds research at the intersection of science and religion, but they're controversial. Some scientists see them as smuggling religion into science; some believers see them as compromising faith with secular assumptions. Yet their funded research often illuminates unexpected connections. They support scholars who might otherwise have no platform for exploring these questions.

Digital Spaces and Virtual Communities

Late at night, when insomnia strikes and cosmic questions feel urgent, I turn to online resources. YouTube has become an unexpected seminary and science classroom combined. PBS Space Time offers exceptional physics education without religious agenda—just clear, accurate explanation of what science actually says versus what popularizers claim. Understanding real cosmology, not simplified versions, is essential for honest dialogue.

The podcast "On Being" occasionally explores science-faith territory with remarkable depth. Krista Tippett interviews physicists and theologians, often in the same episode, creating conversations that would never happen in academic conferences. Listening during my commute has introduced me to thinkers I'd never otherwise encounter.

Online forums provide community for those feeling isolated in their struggle. The BioLogos Forum hosts discussions among Christians accepting evolution. The Strange Notions website creates dialogue between Catholics and atheists. Reddit's various science and religion subreddits offer everything from profound insights to terrible arguments, but the diversity of perspectives prevents echo chambers.

I've learned to approach online resources with caution, though. The internet amplifies both wisdom and foolishness. For every thoughtful exploration of quantum mechanics and consciousness, there are dozens of sites misusing quantum terminology for New Age spirituality. Discernment is essential.

Creating Your Own Library

Building a personal library for science-faith exploration is like preparing for a journey through unknown territory. You need maps from different cartographers, supplies for various weather, tools for unexpected obstacles. No single book or perspective will suffice.

Start with primary sources in both domains. Read Darwin's "Origin of Species," not just books about evolution. Read Augustine's "Confessions," not just summaries of his theology. Primary sources reveal the actual arguments and evidence, not filtered through interpreters with agendas.

Include hostile witnesses—authors who challenge your position forcefully but fairly. If you're a believer, read Dawkins, Dennett, and Harris. If you're a skeptic, read Plantinga, Craig, and McGrath. Not to defeat them but to understand why thoughtful people reach opposite conclusions from similar evidence.

Seek out complexity rather than simplicity. Books that acknowledge difficulty, admit uncertainty, and resist easy answers will serve you better than those claiming to resolve all tensions. The science-faith dialogue is complicated because reality is complicated. Simplistic solutions are usually false solutions.

Pay attention to how books make you feel. The ones that generate anxiety, that threaten cherished beliefs, that make you want to throw them across the room—these might be the most important to engage. Growth happens at the edges of comfort zones.

Practices for the Journey

Beyond reading, certain practices have sustained my exploration of cosmology and faith. Keeping a question journal helps track the evolution of my thinking. Questions I wrote years ago now seem naive; current questions will probably seem equally limited in the future. But documenting the journey helps me see growth and recognize recurring themes.

Regular conversations with those who disagree have proven invaluable. My monthly coffee with a colleague who's a convinced atheist keeps me honest about the strength of materialist arguments. My friendship with a young-earth creationist reminds me that intelligent people can look at the same evidence and reach different conclusions.

Silence and contemplation balance intellectual investigation. Sometimes I need to stop analyzing the cosmic microwave background and simply stand under the stars in wonder. Sometimes I need to cease theological speculation and simply pray, wordlessly, into the mystery.

Warning Signs on the Path

Through years of exploration, I've learned to recognize intellectual and spiritual hazards. Beware of any resource claiming to prove or disprove God through science—both claims exceed what evidence allows. Be suspicious of quantum mysticism that misuses physics terminology for spiritual claims. Avoid conspiracy theories about scientific suppression of evidence for design or religious suppression of evidence against it.

Watch for your own tendencies toward confirmation bias. If you only read authors who agree with you, only engage evidence that supports your position, only have conversations that reinforce your views, you're not exploring but fortifying. Real exploration requires vulnerability to being wrong.

Notice when you're forcing harmony where none exists or creating conflict where none is necessary. Not every scientific discovery challenges faith; not every theological claim has scientific implications. Discerning real tensions from imagined ones requires careful thought.

The Most Important Resource

After all these books, organizations, and online resources, the most important resource remains your own honest reflection. No external authority can resolve these tensions for you. Others can share their journeys, but you must walk your own path.

The questions that keep you awake at night are your curriculum. The doubts that trouble your faith or your materialism are your teachers. The experiences that don't fit your worldview are your data. Your life is the laboratory where cosmology and Christianity meet, and you are both experimenter and experiment.

Regular self-examination helps track your journey. What are you afraid to examine closely? Where are you forcing agreement that doesn't exist? What would you need to change if certain evidence emerged? Are you seeking truth or confirmation? Can you live with permanent uncertainty about ultimate questions?

Equipped for Uncertainty

This resource guide doesn't promise resolution but equips for exploration. The journey between cosmology and Christianity requires intellectual courage to face difficult questions, spiritual stamina to sustain faith amid uncertainty, and emotional resilience to live with unresolved tensions.

Some who embark on this exploration will find integration that satisfies both mind and heart. Others will choose one domain over another, finding they cannot hold both simultaneously. Still others will live permanently in creative tension, finding the dialogue itself meaningful even without resolution. All are legitimate responses to genuine difficulty.

The resources I've shared won't answer all questions but will deepen them. They won't resolve all tensions but will clarify what's actually at stake. They won't provide certainty but might generate wisdom—the kind that comes from wrestling with mystery rather than solving it.

In the end, engaging both modern cosmology and Christian faith honestly requires more than resources—it requires willingness to be changed by what we discover. Whether that change strengthens faith, transforms it, or challenges it to breaking point, the pursuit of truth is worth the risk.

The universe revealed by modern science is stranger than our ancestors imagined. If God exists, God must be stranger too. Resources can guide exploration, but the journey into that strangeness—scientific and spiritual—each person must ultimately make alone, sustained by communities of fellow seekers, guided by whatever light we can find in the cosmic dark.

The path ahead is uncertain, but you're equipped for the journey. Take what serves, leave what doesn't, and above all, keep walking. The universe is vast, mystery is deep, and the conversation between cosmology and faith has barely begun.

Conclusion: Living the Questions

Where We've Traveled

We began this journey standing beneath a night sky, feeling simultaneously dwarfed by cosmic immensity and called by something transcendent. We've explored how modern cosmology challenges traditional Christian faith while potentially deepening it in unexpected ways. Now, as we near the end of our exploration, I find myself back under those same stars, but seeing them differently.

The universe we've discovered through these chapters is one of profound tensions. Prayer seems absurd in a possibly deterministic block universe, yet humans persist in praying across all cultures and centuries. Purpose appears absent from a cosmos trending toward heat death, yet consciousness emerged and seeks meaning with desperate intensity. Divine action seems impossible in law-governed nature, yet believers experience providence too profound for coincidence. Human significance vanishes at cosmic scales, yet we remain the only known moral agents in the observable universe.

We've resisted the temptation to resolve these tensions through clever theology or selective science. The universe often seems more consistent with blind process than loving purpose. Evolution's cruelty challenges belief in benevolent creation. Physical predictions of cosmic death contradict religious hope for transformation. Living with integrity requires acknowledging these difficulties rather than explaining them away with smooth apologetics or defensive fundamentalism.

What We Haven't Resolved

Let me be clear about what this exploration has not accomplished, because honesty about limitations is essential to whatever truth we might have found.

We haven't proven God exists or that Christianity is true. Cosmology offers no such proof. Fine-tuning might suggest design, but it might equally reflect selection effects in a multiverse we can't observe. The Big Bang might resonate with creation ex nihilo, but it might simply describe physical beginnings without metaphysical implications. The resurrection might demonstrate transformation beyond physics, or it might be legendary development of historical events we can't recover.

We haven't shown that science and faith easily harmonize. Despite centuries of dialogue, fundamental tensions remain unresolved and perhaps unresolvable. The universe revealed by telescopes and particle accelerators often seems indifferent to meaning, hostile to life, unconscious of beauty. The God proclaimed by faith seems incompatible with quantum randomness, evolutionary waste, and cosmic emptiness.

We haven't provided a systematic theology adequate to modern cosmology or a scientific framework that incorporates spiritual experience. Both projects might be impossible. Reality might be too complex, too strange, too paradoxical for any single framework to capture. Our minds, evolved for survival on African savannas, might simply lack the capacity to comprehend ultimate reality.

What We Might Have Gained

Despite these limitations—or perhaps because of them—the exploration offers gifts I didn't expect when we began.

We've gained intellectual honesty about the genuine challenges each domain poses to the other. We can engage both cosmology and Christianity without pretending they easily align, without forcing harmony where none exists, without denying tensions that persist despite our best efforts at integration. This honesty might sacrifice comfort, but it gains integrity.

Our imagination has expanded to accommodate a stranger universe and perhaps a greater God. The cosmos is not the cozy dome our ancestors imagined but vast beyond comprehension, old beyond memory, headed toward either transformation or heat death. If God exists, this God must be compatible with quantum mechanics and general relativity, with dark energy and black holes, with the seemingly pointless suffering of evolution. This God cannot be small, manageable, or domesticated.

We've discovered that good questions might be more valuable than poor answers. Instead of settling for shallow reconciliation or defensive separation, we've identified profound questions that deserve continued exploration: How does consciousness emerge from matter? How might eternal God relate to temporal creation? What does moral significance mean at cosmic scales? Living with such questions requires spiritual and intellectual maturity, but it keeps us open to truth we haven't yet discovered.

Perhaps most importantly, we've found we're not alone in this struggle. Scientists around the world wrestle with the same tensions, whether they admit it publicly or not. Believers throughout history have faced challenges to faith from new discoveries about reality. The dialogue between cosmology and Christianity is not our burden alone but a shared human endeavor, continuing across generations, cultures, and disciplines.

The Choice Before Us

Faced with the cosmos revealed by modern science and the claims of Christian faith, we stand at a crossroads. The paths before us lead in different directions, each with its own integrity and cost.

We could abandon faith, concluding that science has rendered religion obsolete. Accept naturalistic materialism as the complete story. Find meaning in temporary human constructs without transcendent grounding. Live courageously in a universe that doesn't care, creating purpose from purposelessness. Many thoughtful people choose this path, and I respect their choice even as I cannot make it myself.

We could abandon science, at least the parts that challenge faith. Reject mainstream cosmology for young-earth creationism or other alternatives that preserve traditional theology. Maintain certainty by dismissing evidence that threatens it. Find security in fundamentalism that admits no questions. This path offers comfort but requires intellectual sacrifice I cannot make.

We could compartmentalize, keeping science and faith in separate spheres that never interact. Be scientists in laboratories and believers in churches, never allowing the two identities to converse. Avoid tension through isolation. This works professionally but creates psychological splitting that feels dishonest.

We could force harmony, insisting science and faith perfectly align if properly understood. Interpret all evidence to support predetermined conclusions. Smooth over difficulties with clever but unconvincing arguments. This satisfies the desire for coherence but requires distorting both science and theology.

Or we could live in tension, accepting that current understanding creates genuine conflicts we cannot resolve. Pursue truth in both domains while acknowledging neither provides the complete picture. Maintain faith and scientific integrity despite unresolved questions. This is the path I've chosen, not because it's comfortable but because it seems most honest.

Practical Implications

How then shall we live, those of us who choose the tension? The cosmic perspective transforms daily existence in ways both subtle and profound.

Practice intellectual humility, recognizing we understand less than we think. About the universe—95% remains dark matter and dark energy we can't explain. About God—infinite mystery exceeds finite comprehension. About consciousness—how matter becomes aware remains opaque. About meaning—why anything matters in a universe heading toward heat death defies easy answer. This humility should shape both scientific research and religious practice.

Embrace provisionality in our beliefs. Hold scientific theories lightly—they might be revised tomorrow. Hold theological doctrines openly—they might need reinterpretation. Truth matters more than consistency with previous positions. When evidence or experience challenges our worldview, we must be willing to change rather than defending outdated ideas.

Seek truth wherever it leads, even to uncomfortable conclusions. Don't predetermine where investigation must arrive. Follow evidence and experience even when they threaten cherished beliefs. Trust that genuine truth, however challenging, is better than comfortable falsehood. This requires courage, but anything less is intellectual and spiritual cowardice.

Maintain hope without certainty. We can hope for new creation without knowing how God might transform physics. We can work for human flourishing without guaranteed ultimate success. We can pray without understanding the mechanism. We can love without cosmic confirmation that love matters. Hope persists not because of evidence but despite its ambiguity.

Act on current understanding while remaining open to new insight. Uncertainty doesn't justify paralysis. We must make choices based on available knowledge—moral choices, professional choices, spiritual choices. Perfect knowledge isn't prerequisite for meaningful action. We act provisionally but genuinely, doing our best with partial understanding.

The Wager Revisited

Pascal wagered on God's existence because potential gain (eternal life) outweighed potential loss (finite pleasures). Modern cosmology transforms this wager into something more profound. We're wagering not just on afterlife but on the nature of reality itself.

Is the universe purposeful or accidental? The fine-tuning of constants, the emergence of consciousness, the existence of beauty that serves no evolutionary purpose—these might point to intention. Or they might be fortunate accidents in vast randomness. The evidence doesn't compel either conclusion.

Is consciousness fundamental or emergent? We experience ourselves as more than neurons, feel meaning that transcends survival, recognize moral truths that contradict genetic interests. This might reveal spiritual reality. Or it might be elaborate illusion created by evolutionary processes. Science cannot yet decide.

Is love cosmically significant or locally constructed? The universe seems indifferent to love, yet love exists, transforming those who experience it. Either love is the universe discovering its purpose through conscious beings, or it's a temporary arrangement of atoms that will dissipate into heat death. Both possibilities require faith to accept.

Is death final or transformable? Physics predicts endings—personal death, species extinction, cosmic heat death. Christianity proclaims transformation—resurrection, new creation, eternal life. These visions seem irreconcilable. We must choose how to live based on which we think more likely, or find a way to hold both possibilities simultaneously.

The wager is existential rather than intellectual. It's about how we choose to live in the face of cosmic ambiguity. If we wager on purpose, consciousness, love, and transformation, we might be projecting meaning onto meaningless cosmos. But if we wager on accident, emergence, construction, and finality, we might be missing reality's deepest dimension. Neither choice is scientifically provable. Both require faith—either in transcendence or its absence.

For Future Explorers

This conversation between cosmology and Christianity has barely begun. Future discoveries will raise new questions we cannot now imagine. Perhaps we'll discover life elsewhere, forcing us to reconceptualize humanity's significance. Perhaps we'll create artificial consciousness, challenging our understanding of soul and spirit. Perhaps we'll understand dark matter and dark energy, revealing dimensions of reality currently hidden. Perhaps we'll develop a theory of quantum gravity, unifying our understanding of the very large and very small.

Each discovery will require theological reflection. Some will challenge faith further; others might unexpectedly support it. The dialogue must continue with each generation bringing fresh perspectives, new questions, different sensitivities. What seems irreconcilable to us might find unexpected resolution. What appears harmonious might reveal hidden tensions.

To future explorers, I offer this testimony: We did our best with available knowledge. We tried to be honest about difficulties while maintaining hope. We refused false choices between science and faith, recognizing both as essential to human flourishing. We lived in tension because reality seemed to demand it.

You will face questions we couldn't ask, have tools we couldn't imagine, see truths we couldn't perceive. But you will also inherit the same fundamental mystery: Why does anything exist? Why does it have the order it has? Why did it produce consciousness capable of contemplating its own existence? These questions transcend scientific epochs. Each generation must engage them anew.

Personal Testament

As we conclude, let me speak personally. Writing this book has been an exercise in intellectual and spiritual honesty that has changed me. I understand better why cosmology challenges faith—the scales are too vast, the processes too impersonal, the future too bleak. I also understand why faith persists despite challenges—human experience includes dimensions that science cannot capture, consciousness exceeds physical description, love and beauty and moral truth seem to point beyond themselves.

Some days, the universe seems obviously purposeful. The elegance of physical laws, the fine-tuning of constants, the emergence of consciousness—surely these point to intention. Other days, it seems obviously accidental. The vast emptiness, the random cruelty, the inevitable heat death—surely these indicate indifference. I've learned to live with this oscillation rather than forcing premature resolution.

I continue both scientific investigation and religious practice not because they harmonize but because abandoning either would diminish my humanity. Science feeds my curiosity about how reality works, my wonder at nature's intricacy, my joy in discovery. Faith nourishes my need for meaning, my hunger for relationship, my hope for transformation. The conversation between them is difficult, sometimes painful, but essential to who I am.

The Last Word That Isn't Last

We end where we began—with wonder at the cosmos and questions about its meaning. Modern cosmology reveals a universe grander and stranger than our ancestors imagined. Christian faith proclaims purposes that transcend physical description. The tension between them might never resolve, but perhaps tension itself is creative, generating energy for continued exploration.

We are stardust that has become conscious, contemplating its origins and destiny. We are the universe becoming aware of itself, asking why it exists, seeking meaning in apparent meaninglessness. We are matter that has developed the capacity for love, beauty, worship—either as cosmic accident or profound purpose.

This is either absurd or amazing. The evidence doesn't compel either conclusion. We must choose how to understand our strange situation—products of 13.8 billion years of cosmic evolution who feel called to transcendence, arrangements of atoms that experience themselves as souls, physical beings who intuit spiritual reality.

The conversation continues. New voices will join, bringing

perspectives we cannot anticipate. The universe keeps expanding, and our understanding expands with it. What seems impossible to reconcile today might find unexpected harmony tomorrow. Or the tension might deepen further, requiring even greater faith to maintain both scientific integrity and religious hope.

Either way, the pursuit of truth—scientific and spiritual—continues. We explore the cosmos through telescopes and particle accelerators, seeking to understand its structure and history. We seek the divine through prayer and practice, hoping for connection with ultimate reality. Sometimes these quests conflict, sometimes they converge, always they challenge us to grow beyond current limitations.

In the vast, expanding, mysterious universe, we small, temporary, conscious beings ask ultimate questions we may never fully answer. The questioning itself might be the most remarkable thing about us—and perhaps the most important. That the universe produced creatures capable of wonder, able to seek truth, willing to live with mystery—this itself is cause for awe.

The universe will end, whether in heat death or divine transformation. But before it does, it produced us—creatures capable of contemplating our own existence, of choosing love despite entropy, of maintaining hope despite evidence. Whether this is absurd accident or amazing grace remains the question each must answer in the solitude of their own cosmic existence.

Choose thoughtfully. Live fully. Question honestly. Hope boldly.

The cosmos awaits our exploration. The Mystery invites our approach. The journey has just begun, and we are all, always, just beginning.

Under the stars tonight, I stand where we started but see differently. The points of light are nuclear furnaces. The spaces between them are expanding. The whole system is heading toward heat death or transformation. And somehow, in this vast, cold, possibly purposeless cosmos, consciousness arose and asks "Why?" That question—our question—echoes through the cosmos, waiting for an answer that may never come, or may have been there all along, hidden in plain sight among the stars.

It is with a sense of shared journey and profound gratitude that I welcome you to explore the life, mind, and enduring legacy of the Apostle Paul through these writings. For me, delving into Paul's world—his dramatic transformation, his complex theology, his tireless ministry, and the powerful letters he left behind—has been a deeply personal and intellectually invigorating pursuit. Paul remains a figure of immense inspiration, a "chosen vessel" whose encounter with the divine fundamentally altered not only his own identity but the course of history.

In my books, whether I am attempting a creative reconstruction like a hypothetical "final epistle" or lost letters, analyzing the psychological depths of his conversion, or seeking to illuminate the core truths within his canonical writings like Romans or the shorter epistles, my aim has consistently been twofold. First, to engage rigorously with the historical context and theological substance of Paul's message. Second, and perhaps more importantly, to bridge the gap between his ancient world and our own, making his insights accessible and demonstrating their relevance for our faith, our communities, and our personal journeys of transformation today.

I do not see these texts merely as academic subjects, but as living words carrying the potential for spiritual growth and a deeper understanding of God's grace. My hope is that as you read, you will be not only informed but also challenged, inspired, and drawn into your own meaningful encounter with the essential Paul. Thank you for joining me in this exploration. May it enrich your understanding and encourage you on your own path of faith.

Sincerely, Arthur A. Tiger

arthurtigerbooks.com

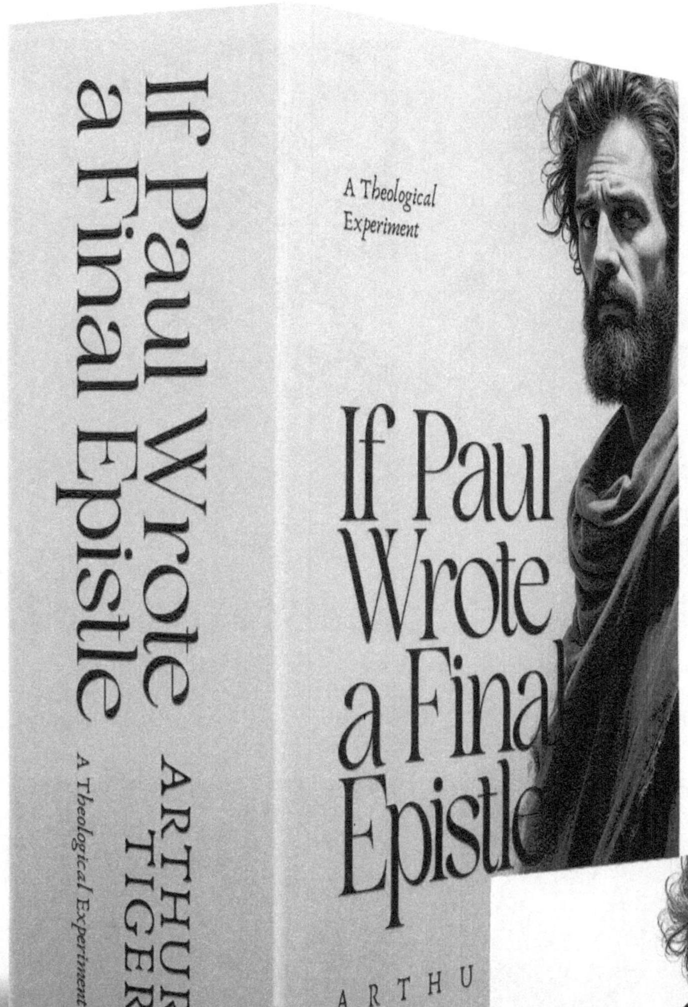

Academic work for seminary students. A theological experiment revealing Paul's theology, imagining Paul's final letter to all Christians, grounded in his authentic writings and exploring faith, unity, and practical Christian living. And based on the scholarly study "If Paul Wrote a Final Epistle," this accessible adaptation brings this innovative concept to a wider audience. Every word comes directly from Paul's authentic New Testament letters, woven together to address the universal church throughout history.

The book explores human existence, consciousness, and free will through modern physics and philosophy.

Dive into the world of dinosaurs and discover how science and faith intersect in an attempt to unravel the mysteries of our planet's past.

Human knowledge is limited, opening the way to faith and understanding realities beyond the senses.

The book connects Christian faith with extraterrestrial life, asserting that science and faith are not contradictory.

The book explores escapism and how to balance healthy rest with avoiding reality.

The book is a philosophical manifesto about life's harsh rules, shattering illusions and revealing uncomfortable truths.

This guided prayer journal helps beginners develop a meaningful prayer life through conversation with God.

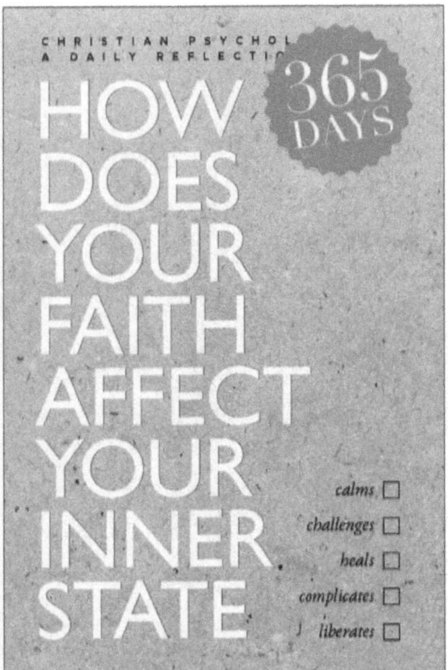

In a world that demands constant productivity, this journal offers another way — one that honors both your faith and your human limits.

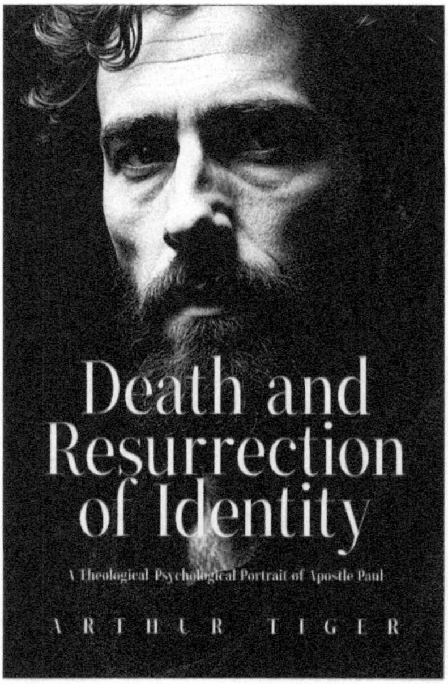

The book is a theological-psychological portrait of the Apostle Paul, exploring his radical personality transformation.

Starting with a provocative declaration—"There Is No God!"—this book doesn't end where most skeptical explorations do.

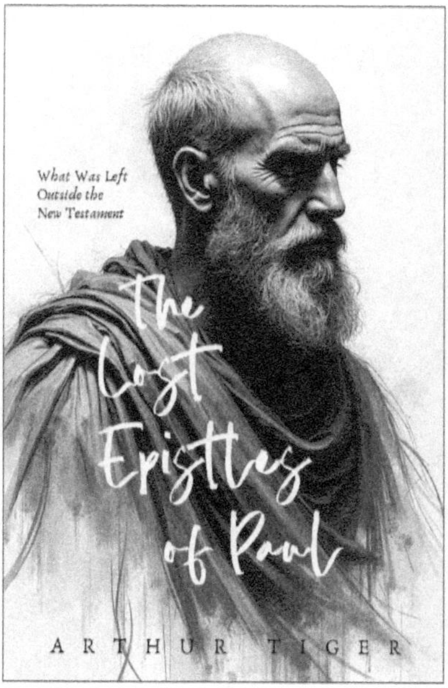

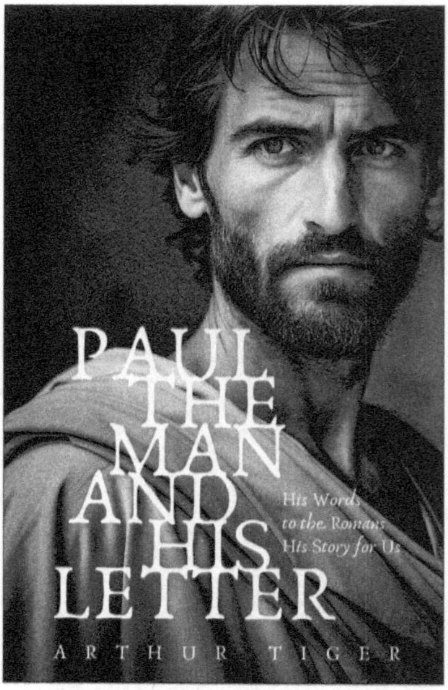

The Bible mentions Paul's letters that vanished. What secrets did they hold? Uncover the missing pieces of Christianity's greatest theologian.

This book reflects on Paul's letter to the Romans, exploring how his words can transform lives and deepen understanding.

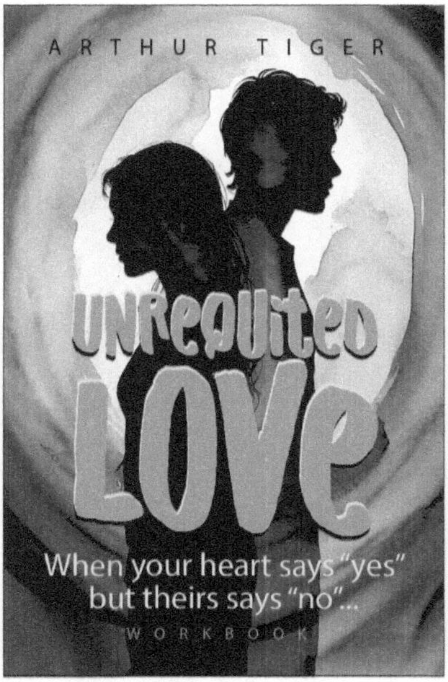

One-sided love is a crushing weight. Hope flares up with every interaction, only to be extinguished by the painful truth that they don't feel the same.

Cracks That Let the Light In is a deeply heartfelt journey into the nature of healing, self-acceptance, and transformation.

A practical guide to building a strong and trusting relationship.

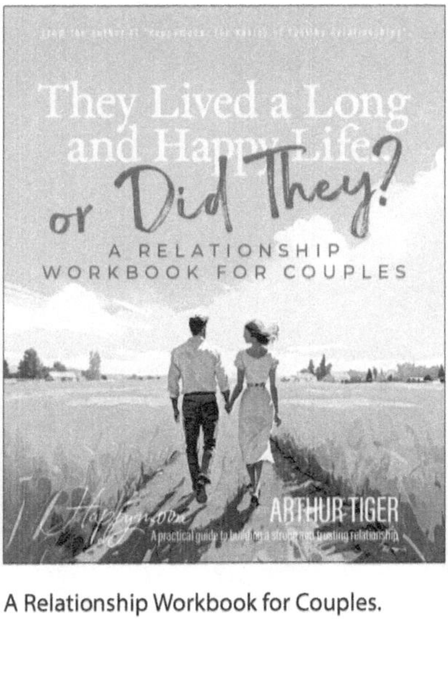

A Relationship Workbook for Couples.

www.ingramcontent.com/pod-product-compliance
Lightning Source LLC
Chambersburg PA
CBHW031349230926
51848CB00002B/448